Magnesium and Its Alloys as Implant Materials

Magnesium and Its Alloys as Implant Materials

Corrosion, Mechanical and Biological Performances

Mirco Peron
Filippo Berto
Jan Torgersen

CRC Press
Taylor & Francis Group
Boca Raton London New York

CRC Press is an imprint of the
Taylor & Francis Group, an **informa** business

First edition published 2020
by CRC Press
6000 Broken Sound Parkway NW, Suite 300, Boca Raton, FL 33487-2742

and by CRC Press
4 Park Square, Milton Park, Abingdon, Oxon OX14 4RN

First issued in paperback 2023

© 2020 Taylor & Francis Group, LLC

CRC Press is an imprint of Taylor & Francis Group, an Informa business

ISBN-13: 978-0-367-42945-4 (hbk)
ISBN-13: 978-1-03-265433-1 (pbk)
ISBN-13: 978-1-00-300032-7 (ebk)

DOI: 10.1201/9781003000327

Contents

List of Abbreviations

AE	Mg alloy with main alloying elements aluminum and rare earth
Ag	Silver
Al	Aluminum
AM	Mg alloy with main alloying elements aluminum and manganese
APD	Accumulative plastic deformed
ARB	Accumulative roll-bonded
ARR	asymmetric reduction rolled
AU	Artificial urine
AZ	Mg alloy with main alloying elements aluminum and zinc
B	Boron
BDF	Bi-directional forged
BE	Backward extruded
Bi	Bismuth
BMGs	Bulk metallic glasses
BP	Back pressure
Ca	Calcium
Ce	Cerium
CEC	Cyclic extrusion and compression
CEE	Cyclic expansion extrusion
CF	Corrosion fatigue
CP	Commercially pure Mg
CR	Cross rolling
Cr	Chromium
Cu	Copper
DC	direct contact
DCC	direct chilling cast
DCE	Double continuous extrusion
DCT	Deep cryogenically treated
DMD	Disintegrated melt deposition
DMEM	Dulbecco's modified Eagle's medium
DSR	Differential speed rolled
DSST	Double step solution treated
EBSD	Electron backscatter diffraction
EBSS	Earle's balanced salt solution
ECAE	Equal channel angular extruded
ECAP	Equal channel angular pressed
ECAP-BP	Equal-channel angular pressed with applied back pressure
ECM	Endothelial cell medium
E_{corr}	Corrosion potential
EPT	Electropulsing treatment
ESR	Equal speed rolling

FBS	Fetal bovine serum
Fe	Iron
FSP	Friction stir processed
Gd	Gadolinium
GPMC	Gravity permanent mold casted
Hank's	Hank's solution
HBSS	Hank's balanced salt solution
HCl	Hydrochloric acid
HE	Hydrostatic extruded
HEMA	High-energy mechanical alloyed
HP	High-pure Mg
HPDC	High-pressure die-casted
HPT	High-pressure torsion
HRDSR	high-ratio differential speed rolling
HSRMF	High strain rate multiple forging
HSRR	High strain rate rolled
HTP	H-tube pressed
IC	indirect contact
I_{corr}	Corrosion current
i_{corr}	Corrosion current density
IR	Isothermal rolling
KBM	Kirkland's biocorrosion medium
La	Lanthanum
LAE	Mg alloy with main alloying elements lithium, aluminum and rare earth
LAM	Laser additive manufacturing
LB	Lysogeny broth
LENS	Laser-engineered net shaping
Li	Lithium
LPDC	Low pressure die-casted
LSP	Laser shock peening
LSRMF	Low strain rate multiple forging
LTUSSE	Low temperature ultra-slow-speed extruded
MAD	multiaxial deformation
MAF	Multidirectional forged
MC-HPDC	High pressure die-casted with melt conditioning
MDF	Multi-direction forged
MEM	Minimum essential medium
MEMp	MEM containing 40 g/L bovine serum albumin
MIF	Multiaxial isothermal forged
MM	Misch metal
Mn	Manganese
MPF	Multielectrolyte physiological fluid
MPHR	Multi-pass hot rolled
MTE	Microtube extrusion
Nb	Niobium

Nd	Neodymium
Ni	Nickel
PBS	Phosphate-buffered saline
PEO	Plasma electrolytic oxidation
PM	Powder metallurgy
PMF	Pulse magnetic field
Pr	Praseodymium
PVD	Physical vapor deposition
RAP	Recrystallization and partial melting
RB	Reverse bending
RDC	Rheo-diecasted
RE	Rare earth
REE	Rare-earth elements
RHT	Recrystallization heat treatment
Ringer's	Ringer's solution
RS	Rapid solidified
RSC	Rheo-squeeze casting
RS P/M	Rapid solidification powder metallurgy
RTE	Room temperature extruded
RU	Repeated upsetting
Sb	Antimony
SBF	Simulation body fluid
Sc	Scandium
SCC	Stress corrosion cracking
SHT	Solution heat treated
Si	Silicon
SLM	Selective laser melted
SMAT	Surface mechanical attrition treated
SMCM	Smooth muscle cell medium
SMF	Steady magnetic field
Sn	Tin
SPD	Severe plastic deformation
SR	Symmetrical rolled
Sr	Strontium
S-RS P/M	Simplified rapid solidification powder metallurgy
SSP	Severe shot peening
SSST	Single step solution treated
SSTT	Semi-solid thermal transformation
SVDC	Super vacuum die-casted
Ti	Titanium
TRC	Twin roll casted
TSB	Tryptic soy broth
UHP	Ultra-high-pure Mg
UR	Unidirectional rolling
UTS	Ultimate tensile strength
UV	Ultrasonic vibration

V	Vanadium
VFUT	Variable frequency ultrasonic treatment
WE	Mg alloy with main alloying elements yttrium and rare-earth element
XHP	Ultra-high purity
Y	Yttrium
ZE	Mg alloy with main alloying elements zinc and rare-earth element
ZK	Mg alloy with main alloying elements zinc and zirconium
ZM	Mg alloy with main alloying elements zinc and manganese
Zn	Zinc
ZW	Mg alloy with main alloying elements zinc and yttrium
ZX	Mg alloy with main alloying elements zinc and calcium
αMEM	Alpha minimum essential medium

Authors

Mirco Peron earned his degree in mechanical engineering (summa cum laude) in 2015 from the University of Padova, where his thesis evaluated the fatigue damage and stiffness evolution in composite laminates. He is currently a PhD student at Norwegian University of Science and Technology (NTNU), Trondheim. His PhD topic deals with the optimization of mechanical and corrosion properties of magnesium and its alloys for biomedical applications, with particular reference to the corrosion-assisted cracking phenomena.

Filippo Berto is Chair of Structural Integrity at the Norwegian University of Science and Technology in Norway. He is in charge of the Mechanical and Material Characterization Lab in the Department of Mechanical and Industrial Engineering. He is author of more than 500 technical papers, mainly oriented to materials science engineering, the brittle failure of different materials, notch effect, the application of the finite element method to the structural analysis, the mechanical behavior of metallic materials, the fatigue performance of notched components as well as the reliability of welded, bolted and bonded joints. Since 2003, he has been working on different aspects of the structural integrity discipline, by mainly focusing attention on problems related to the static and fatigue assessment of engineering materials with particular attention to biomedical and medical applications and materials.

Jan Torgersen is Professor of mechanical engineering at NTNU, Trondheim. He received his PhD from Vienna University of Technology, where he worked on high-resolution laser microfabrication of hydrogels for tissue engineering. He was pioneering in the work of processing hydrogel formulations at micron scale resolution in vivo, in the presence of living cells and whole organisms. He received a postdoctoral fellowship to work on a nanoscale vapor deposition technique called atomic layer deposition, allowing conformal coating of thermally fragile and nanostructured substrates with atomically thin layers of a wide range of materials. He contributed to the development of a self-limiting deposition process for high-k materials for Dynamic Random Access Memory (DRAM) applications. His current research interests are micro- and nanofabrication as well as surface functionalization, with particular focus on biomedical applications.

1 Introduction

1.1 INTRODUCTION

Biomedical implants have played a fundamental role in improving people's health worldwide. They are used in applications such as orthopedics, cardiovascular stent and neural prosthetics, where there is abundant need to replace or repair fractured or diseased parts of the human body [1–4]. Among these, orthopedic surgery is characterized by the highest annual growth rate [3]. According to Long and Rach [5], almost 90% of the population over 40 years is affected by degenerative joint diseases. Total hip replacements are predicted to represent half of the estimated total number of operations in 2030 [6]. The surgical implantation of artificial biomaterials of specific size and shape is, in fact, an effective solution in restoring the load-bearing capacity of damaged bone tissue. Although outstanding mechanical and structural properties characterize bones, fractures can happen because of [7–9]

- stresses arising from daily activities
- sudden injuries
- bone infections and tumors, resulting in pathological fractures

Depending on their applications, implant devices can be classified into permanent or temporary. The former is required in applications such as joint replacements where a long-term existence in the human body is required. Appropriate materials are metals [10], polymers [11,12], ceramics [13,14] and composite materials [7,15], and a synthesis of their application has been recently published by the authors [16]. Polymers, such as PEEK (polyetheretherketone), are used as spinal cages, where the stress requirements are not high enough to require the use of metallic materials. In addition, van Dijk et al. [17] reported that polymeric cages significantly reduce the stress shielding phenomenon encountered with the application of titanium cages. Ceramic materials, such as Zirconia (ZrO_2), are used in dental applications due to their high strength (*in vitro* studies reported a flexural strength of 900–1200 MPa [18]), radiopacity and better esthetic [19,20]. Composite materials, such as carbon fiber-reinforced PEEK, have been studied for orthopedic applications such as hip-joint due to the possibility of tailoring their mechanical properties to those of human bones [21]. However, their long-term durability properties are still insufficient and thus metals are preferred [22]. Metallic materials are, in fact, generally used in load-bearing applications, where their high strength and fracture toughness render them preferable to ceramics (due to their higher fracture toughness), polymers (due to their high strength) and composite materials. However, metallic implants have two main problems. First, their elastic modulus highly differs from that of bone: for both stainless steel and cobalt–chrome alloys it is ten times higher, while for Ti-6Al-4V it is five times (Table 1.1).

This leads to the occurrence of the stress shielding phenomenon, which is a phenomenological consequence of stress distribution changes that leads to the progressive bone desorption, phenomenon well described in ref. [16] and extensively reported in literature [28–37]. The second problem stems from their nonbiodegradability that leads to long-term complications, such as local inflammations due to the potential release of cytotoxic ions as a consequence of corrosion and/or wear processes [38–42]. Owing to these causes, the maximum service period of the permanent implant is around 12–15 years [43].

In contrast, temporary devices require biomaterials to stay inside the human body only for a restricted period, that is, as long as a bone heals (3–4 months [44,45]). Biodegradable materials have thus emerged to a greater extent in the fields of engineering scaffolds and bone fixators such as bone plates, screws, pins and stents, where materials that ideally degrade in the same manner and speed as natural bone heals are widely claimed [46–48]. Both natural and synthetic polymers have been studied extensively as biodegradable materials. Natural polymers such as polysaccharides and collagen have all produced favorable outcomes in tissue engineering applications [49–55], while synthetic polymers such as polyglycolic acid, poly-L-lactic acid, poly-DL-lactic acid and poly-capro lactone have been used as biodegradable sutures, drug delivery systems, fixation devices and low-load-bearing applications [56–63]. However, because of their low mechanical strength when compared to metals, polymers have been mostly used in soft tissue reconstruction and low-load-bearing applications. Moreover, they might absorb liquids and swell, leach undesirable products such as monomers, fillers and antioxidants. Furthermore, the sterilization process might affect their properties [7]. The combination of high strength and biocompatibility can be found in biodegradable metal alloys. Several of them, such as iron-based metals, Zn-based metals

TABLE 1.1

Comparison of the mechanical properties of natural bone with various implant materials [10,15,23–27]

Properties	Natural bone	Stainless steel	Ti alloy	Co–Cr alloy	Magnesium
Density (g/cm^3)	1.7–2.0	7.9–8.1	4.4–4.5	8.3–9.2	1.74–2.0
Elastic modulus (MPa)	3–20	189–205	110–117	230	41–45
Tensile strength (MPa)	80–150	480–620	930–1140	900–1540	170–270
Compressive yield strength (MPa)	130–180	170–310	758–1117	450–1000	65–100
Elongation at failure (%)	1–7	30–40	8–15	30–45	6–20
Fracture toughness (MPa m$^{1/2}$)	3–6	50–200	55–115	100	15–40

and tungsten, have been studied [64], but most of the scientific efforts focus on magnesium and its alloys [64–66]. Among metallic engineering materials, magnesium possesses, in fact, one of the best biocompatibility with human physiology and the best mechanical compatibility with human bones [67]. The density of magnesium and its alloys (1.74–2 g/cm^3) is, in fact, very similar to that of cortical bone (1.7–2 g/cm^3). In addition, the similarity of the elastic modulus of magnesium and its alloys with that of natural bones (Table 1.1) potentially reduces the possibilities of stress shielding in hard tissue applications [16,68]. Moreover, magnesium is the fourth most abundant element in the human body: the human body usually contains 35 g of magnesium per 70 kg of body weight, and it is recommended that an adult receives 240–420 mg daily [27]. It is a cofactor for many enzymes (it is involved in more than 300 enzymatic reactions), and it has roles in protein and nucleic acid synthesis, mitochondrial activity and integrity and in many other cellular functions [69–72]. Finally, Mg^{2+} ions, resulting from the degradation process (see Chapter 2), are reported to aid the healing and growth of tissue. Any excess of these ions is harmlessly excreted in the urine [24,73]. However, despite its many advantages, magnesium has some disadvantages. First, the strength of cast pure magnesium is too low compared to that of human bones [74]. This issue has been overcome by either alloying magnesium or reducing its grain size through thermo-mechanical processes (extrusion, rolling and so on) or severe plastic deformation techniques (ECAP, MAF and so on). Another disadvantage of magnesium and its alloys is their high corrosion rate in the body that may lead to a loss of mechanical integrity before tissues have sufficient time to heal. Moreover, hydrogen as a corrosion product together with the generation of respective hydrogen pockets can influence the healing process or, if the pockets are large, they may cause the death of patients through blocking of the blood stream [75,76]. Finally, the simultaneous action of the corrosive human body fluid and the mechanical loading can cause further complications of sudden fracture of implants due to corrosion-assisted cracking, such as stress corrosion cracking and corrosion fatigue [77,78].

Despite their high potential, magnesium and its alloys are not yet being utilized in biomedical applications, as the aforementioned challenges could yet not be overcome. In this work, we aim to provide scientific insights into these challenges and give a thorough state-of-the-art overview of the research conducted thus far, particularly focusing on the mechanical properties, corrosion behavior and biological performances of these materials. The majority of the alloys are initially introduced for industrial applications and then evaluated for biomedical applications. Here, the main alloys are reviewed with regard to their suitability for biomedical applications. Hence, their mechanical properties (tensile and compressive yield and ultimate strength, tensile elongation to failure), corrosion properties in terms of electrochemical properties (E_{cor} and i_{cor}) and corrosion rates and their biocompatibility expressed in terms of cell viability and hemolysis rate are in focus. Reviewing more than 1200 papers, the authors aim to provide an extensive database of these properties taking a critical look at the bio-performances. Previous reviews on biocompatible Mg alloys gave an extensive view on corrosion [43,68], with some papers also looking at mechanical [44] and biocompatible [73] properties. Temporary implantation, however, challenges all of these aspects. In this work, we hence aim to link these aspects, reporting the main challenges and common strategies and their interplay with respect to these properties. In particular, the effect that each strategy employed to improve a certain property has on the other properties will be considered. Going through this work, the authors intend to put the reader into the position to accurately discern the proper Mg-based material for his/her applications and to choose the proper improvement strategy for his/her cause. To this aim, the book is structured as follows: in Chapter 2, the main challenges hampering the use of magnesium in biomedical applications and the common improvement strategies are listed. In Chapter 3, the most investigated Mg alloys are reported in separate sections, detailing their mechanical properties, corrosion behavior and biotoxicity. High-pure and ultra-high-pure Mg, Al-based Mg alloys, Zn-based Mg alloys, Ca-based alloys and rare earth-based Mg alloys have been considered. In Chapter 4, the performances of the alloys with respect to the challenges are summarized, providing the reader with useful information and suggestions on the potentially most suited choice. Finally, in Chapter 5, an outlook portraying the authors' opinion of the future development of the field is provided. Readers first approaching the topic can get familiar with the Mg corrosion process, the nonuniform corrosion modes and testing procedures to evaluate the corrosion behavior by reading Appendix A. Common testing procedures to evaluate biocompatibility are briefly described in Appendix B.

REFERENCES

[1] Regar, E., Sianos, G., Serruys, P.W. (2001). Stent development and local drug delivery, *Br. Med. Bull.*, 59, pp. 227–48.

[2] Greatbatch, W., Holmes, C.F. (1991). History of implantable devices, *IEEE Eng. Med. Biol. Mag.*, 10(3), pp. 38–41, Doi: 10.1109/51.84185.

[3] Long, P.H. (2008). Medical devices in orthopedic applications, *Toxicol. Pathol.*, 36 (1), pp. 85–91, Doi: 10.1177/0192623307310951.

[4] Khan, W., Muntimadugu, E., Jaffe, M., Domb, A.J. (2014). *Implantable medical devices*, Eds. A.J. Domb, W. Khan, Boston, MA, Springer US, pp. 33–59.

[5] Long, M., Rack, H. (1998). Titanium alloys in total joint replacement – A materials science perspective, *Biomaterials*, 19(18), pp. 1621–39, Doi: 10.1016/S0142-9612(97)00146-4.

[6] Gultepe, E., Nagesha, D., Sridhar, S., Amiji, M. (2010). Nanoporous inorganic membranes or coatings for sustained drug delivery in implantable devices, *Adv. Drug Deliv. Rev.*, 62(3), pp. 305–15, Doi: 10.1016/j.addr.2009.11.003.

[7] Ramakrishna, S., Mayer, J., Wintermantel, E., Leong, K.W. (2001). Biomedical applications of polymer-composite materials: A review, *Compos. Sci. Technol.*, 61 (9), pp. 1189–224, Doi: 10.1016/S0266-3538(00)00241-4.

[8] Black, J. (2006). *Biological performance of materials: Fundamentals of biocompatibility*, Boca Raton, FL, CRC Taylor & Francis.

[9] Hamblen, D.L., Simpson, A.H.R.W., Adams, J.C., Adams, J.C. (2007). *Adams's outline of fractures, including joint injuries*, London, UK, Churchill Livingstone Elsevier.

[10] Hanawa, T. (2010). Overview of metals and applications. In *Metals for biomedical devices*, Ed. M. Niinomi, Cambridge, UK, Woodhead Publishing Elsevier, pp. 3–24.

[11] Lendlein, A., Rehahn, M., Buchmeiser, M.R., Haag, R. (2010). Polymers in biomedicine and electronics, *Macromol. Rapid Commun.*, 31(17), pp. 1487–91, Doi: 10.1002/marc.201000426.

[12] Pruitt, L., Furmanski, J. (n.d.). Polymeric biomaterials for load-bearing medical devices, *JOM*, 61(9), pp. 14–20, Doi: 10.1007/s11837-009-0126-3.

[13] Dubok, V.A. (2000). Bioceramics – Yesterday, today, tomorrow, *Powder Metall. Met. Ceram.*, 39(7/8), pp. 381–94, Doi: 10.1023/A:1026617607548.

[14] Hench, L.L. (1991). Bioceramics: From concept to clinic, *J. Am. Ceram. Soc.*, 74 (7), pp. 1487–510, Doi: 10.1111/J.1151-2916.1991.TB07132.X.

[15] Wang, M. (2004). *Bioactive ceramic-polymer composites for tissue replacement*, Singapore, World Scientific, pp. 8-1–8-29.

[16] Peron, M., Torgersen, J., Berto, F. (2017). Mg and its alloys for biomedical applications: Exploring corrosion and its interplay with mechanical failure, *Metals* (Basel), 7(7), p. 252, Doi: 10.3390/met7070252.

[17] van Dijk, M., Smit, T.H., Sugihara, S., Burger, E.H., Wuisman, P.I. (2002). The effect of cage stiffness on the rate of lumbar interbody fusion: An in vivo model using poly (l-lactic Acid) and titanium cages, *Spine*, (Phila. Pa. 1976) 27(7), pp. 682–8.

[18] Tinschert, J., Zwez, D., Marx, R., Anusavice, K.J. (2000). Structural reliability of alumina-, feldspar-, leucite-, mica- and zirconia-based ceramics, *J. Dent.*, 28 (7), pp. 529–35, Doi: 10.1016/S0300-5712(00)00030-0.

[19] Della Bona, A., Pecho, O.E., Alessandretti, R. (2015). Zirconia as a dental biomaterial, *Materials*, (Basel, Switzerland) 8(8), pp. 4978–91, Doi: 10.3390/ma8084978.

[20] Daou, E.E. (2014). The zirconia ceramic: Strengths and weaknesses, *Open Dent. J.*, 8, pp. 33–42, Doi: 10.2174/1874210601408010033.

[21] Williams, D.F., McNamara, A., Turner, R.M. (1987). Potential of polyetheretherketone (PEEK) and carbon-fibre-reinforced PEEK in medical applications, *J. Mater. Sci. Lett.*, 6(2), pp. 188–90, Doi: 10.1007/BF01728981.

[22] Kutz, M. (2004) *Standard Handbook of Biomedical Engineering & Design*, New York, McGraw-Hill, pp. 12.1–12.17.

[23] Zhang, L.-N., Hou, Z.-T., Ye, X., Xu, Z.-B., Bai, X.-L., Shang, P. (2013). The effect of selected alloying element additions on properties of Mg-based alloy as bioimplants: A literature review, *Front. Mater. Sci.*, 7(3), pp. 227–36, Doi: 10.1007/s11706-013-0210-z.

[24] Staiger, M.P., Pietak, A.M., Huadmai, J., Dias, G. (2006). Magnesium and its alloys as orthopedic biomaterials: A review, *Biomaterials*, 27, pp. 1728–34, Doi: 10.1016/j.biomaterials.2005.10.003.

[25] DeGarmo, E.P., Black, J.T., Kohser, R.A. (2011). *DeGarmo's materials and processes in manufacturing*, Danvers, MA, John Wiley & Sons, p. 1184, Doi: 10.1017/CBO9781107415324.004.

[26] Gibson, L.J., Ashby, M.F. (1988). *Cellular solids: Structure and properties*, Oxford, Pergamon Press.

[27] Hänzi, A.C., Sologubenko, A.S., Uggowitzer, P.J. (2009). Design strategy for new biodegradable Mg–Y–Zn alloys for medical applications, *Int. J. Mater. Res.*, 100 (8), pp. 1127–36, Doi: 10.3139/146.110157.

[28] Bauer, T.W., Schils, J. (1999). The pathology of total joint arthroplasty. II. Mechanisms of implant failure, *Skeletal Radiol.*, 28(9), pp. 483–97.

[29] Dujovne, A.R., Bobyn, J.D., Krygier, J.J., Miller, J.E., Brooks, C.E. (1993). Mechanical compatibility of noncemented hip prostheses with the human femur, *J. Arthroplasty*, 8(1), pp. 7–22, Doi: 10.1016/S0883-5403(06)80102-6.

[30] Engh, C., Bobyn, J., Glassman, A. (1987). Porous-coated hip replacement. The factors governing bone ingrowth, stress shielding, and clinical results, *Bone Joint J.*, 69-B(1), pp. 45–55.

[31] Engh, C.A., Bobyn, J.D. (1988). The influence of stem size and extent of porous coating on femoral bone resorption after primary cementless hip arthroplasty, *Clin. Orthop. Relat. Res.*, 231, pp. 7–28.

[32] Kerner, J., Huiskes, R., van Lenthe, G.H., Weinans, H., van Rietbergen, B., Engh, C.A., Amis, A.A. (1999). Correlation between pre-operative periprosthetic bone density and post-operative bone loss in THA can be explained by strain-adaptive remodelling, *J. Biomech.*, 32(7), pp. 695–703, Doi: 10.1016/S0021-9290(99)00041-X.

[33] Sumner, D.R., Galante, J.O. (1992). Determinants of stress shielding: Design versus materials versus interface, *Clin. Orthop. Relat. Res.*, 274, pp. 202–12.

[34] Sumner, D.R., Turner, T.M., Igloria, R., Urban, R.M., Galante, J.O., Simon, B.R., Gomez, M.A. (1998). Functional adaptation and ingrowth of bone vary as a function of hip implant stiffness, *J. Biomech.*, 31(10), pp. 909–17, Doi: 10.1016/S0021-9290(98)00096-7.

[35] Turner, T.M., Sumner, D.R., Urban, R.M., Igloria, R., Galante, J.O. (1997). Maintenance of proximal cortical bone with use of a less stiff femoral component in hemiarthroplasty of the hip without cement. An investigation in a canine model at six months and two years, *J. Bone Joint Surg. Am.*, 79(9), pp. 1381–90.

[36] Van Rietbergen, B., Huiskes, R., Weinans, H., Sumner, D.R., Turner, T.M., Galante, J.O. (1993). The mechanism of bone remodeling and resorption around press-fitted THA stems, *J. Biomech.*, 26(4–5), pp. 369–82, Doi: 10.1016/0021-9290(93)90001-U.

[37] Wolff, J. (1986). *The law of bone remodelling*, Berlin, Heidelberg, Springer Berlin Heidelberg.

[38] Pound, B.G. (2014). Corrosion behavior of metallic materials in biomedical applications. I. Ti and its alloys, *Corr. Rev.*, 32, pp. 1–20, Doi: 10.1515/corrrev-2014-0007.

[39] Pound, B.G. (2014). Corrosion behavior of metallic materials in biomedical applications. II. Stainless steels and Co-Cr alloys, *Corr. Rev.*, 32, pp. 21–41, Doi: 10.1515/corrrev-2014-0008.

[40] Jacobs, J.J., Gilbert, J.L., Urban, R.M. (1998). Corrosion of metal orthopaedic implants, *J. Bone Joint Surg. Am.*, 80(2), pp. 268–82.

[41] Jacobs, J.J., Hallab, N.J., Skipor, A.K., Urban, R.M. (2003). Metal degradation products: A cause for concern in metal-metal bearings? *Clin. Orthop. Relat. Res.*, 417, pp. 139–47, Doi: 10.1097/01.blo.0000096810.78689.62.

[42] Beech, I.B., Sunner, J.A., Arciola, C.R., Cristiani, P. (2006). Microbially-influenced corrosion: Damage to prostheses, delight for bacteria, *Int. J. Artif. Organs*, 29(4), pp. 443–52.

[43] Manivasagam, G., Suwas, S. (2014). Biodegradable Mg and Mg based alloys for biomedical implants, *Mater. Sci. Technol.*, 30(5), pp. 515–20, Doi: 10.1179/1743284713Y.0000000500.

[44] Li, N., Zheng, Y. (2013). Novel magnesium alloys developed for biomedical application: A review, *J. Mater. Sci. Technol.*, 29(6), pp. 489–502, Doi: 10.1016/J.JMST.2013.02.005.

[45] Witte, F., Kaese, V., Haferkamp, H., Switzer, E., Meyer-Lindenberg, A., Wirth, C.J., Windhagen, H. (2005). In vivo corrosion of four magnesium alloys and the

associated bone response, *Biomaterials*, 26(17), pp. 3557–63, Doi: 10.1016/j.biomaterials.2004.09.049.

[46] Schinhammer, M., Hänzi, A.C., Löffler, J.F., Uggowitzer, P.J. (2010). Design strategy for biodegradable Fe-based alloys for medical applications, *Acta Biomater.*, 6 (5), pp. 1705–13, Doi: 10.1016/j.actbio.2009.07.039.

[47] Vojtěch, D., Kubásek, J., Šerák, J., Novák, P. (2011). Mechanical and corrosion properties of newly developed biodegradable Zn-based alloys for bone fixation, *Acta Biomater.*, 7(9), pp. 3515–22, Doi: 10.1016/j.actbio.2011.05.008.

[48] Kroeze, R.J., Helder, M.N., Govaert, L.E., Smit, T.H. (2009). Biodegradable polymers in bone tissue engineering, *Materials* (Basel), 2(3), pp. 833–56, Doi: 10.3390/ma2030833.

[49] Yang, L., Zhang, L.-M. (2009). Chemical structural and chain conformational characterization of some bioactive polysaccharides isolated from natural sources, *Carbohydr. Polym.*, 76(3), pp. 349–61, Doi: 10.1016/J.CARBPOL.2008.12.015.

[50] Sinha, V.R., Kumria, R. (2001). Polysaccharides in colon-specific drug delivery, *Int. J. Pharm.*, 224(1–2), pp. 19–38, Doi: 10.1016/S0378-5173(01)00720-7.

[51] Sun, R., Fang, J.M., Goodwin, A., Lawther, J.M., Bolton, A.J. (1998). Fractionation and characterization of polysaccharides from abaca fibre, *Carbohydr. Polym.*, 37(4), pp. 351–9, Doi: 10.1016/S0144-8617(98)00046-0.

[52] Crini, G. (2005). Recent developments in polysaccharide-based materials used as adsorbents in wastewater treatment, *Prog. Polym. Sci.*, 30(1), pp. 38–70, Doi: 10.1016/J.PROGPOLYMSCI.2004.11.002.

[53] Sato, T., Chen, G., Ushida, T., Ishii, T., Ochiai, N., Tateishi, T., Tanaka, J. (2004). Evaluation of PLLA–collagen hybrid sponge as a scaffold for cartilage tissue engineering, *Mater. Sci. Eng. C*, 24(3), pp. 365–72, Doi: 10.1016/J.MSEC.2003.12.010.

[54] Dai, N.-T., Williamson, M.R., Khammo, N., Adams, E.F., Coombes, A.G.A. (2004). Composite cell support membranes based on collagen and polycaprolactone for tissue engineering of skin, *Biomaterials*, 25(18), pp. 4263–71, Doi: 10.1016/J.BIOMATERIALS.2003.11.022.

[55] Chun, T.-H., Hotary, K.B., Sabeh, F., Saltiel, A.R., Allen, E.D., Weiss, S.J. (2006). A pericellular collagenase directs the 3-dimensional development of white adipose tissue, *Cell*, 125(3), pp. 577–91, Doi: 10.1016/j.cell.2006.02.050.

[56] Zheng, X., Kan, B., Gou, M., Fu, S., Zhang, J., Men, K., Chen, L., Luo, F., Zhao, Y., Zhao, X., Wei, Y., Qian, Z. (2010). Preparation of MPEG–PLA nanoparticle for honokiol delivery in vitro, *Int. J. Pharm.*, 386(1–2), pp. 262–7, Doi: 10.1016/J.IJPHARM.2009.11.014.

[57] Dong, Y., Feng, S.-S. (2004). Methoxy poly(ethylene glycol)-poly(lactide) (MPEG-PLA) nanoparticles for controlled delivery of anticancer drugs, *Biomaterials*, 25(14), pp. 2843–9, Doi: 10.1016/j.biomaterials.2003.09.055.

[58] Kanczler, J.M., Ginty, P.J., Barry, J.J.A., Clarke, N.M.P., Howdle, S.M., Shakesheff, K.M., Oreffo, R.O.C. (2008). The effect of mesenchymal populations and vascular endothelial growth factor delivered from biodegradable polymer scaffolds on bone formation, *Biomaterials*, 29(12), pp. 1892–900, Doi: 10.1016/j.biomaterials.2007.12.031.

[59] Rimondini, L., Nicoli-Aldini, N., Fini, M., Guzzardella, G., Tschon, M., Giardino, R. (2005). In vivo experimental study on bone regeneration in critical bone defects using an injectable biodegradable PLA/PGA copolymer, *Oral Surg. Oral Med. Oral Pathol. Oral Radiol. Endod.*, 99(2), pp. 148–54, Doi: 10.1016/J.TRIPLEO.2004.05.010.

[60] Rezgui, F., Swistek, M., Hiver, J.M., G'Sell, C., Sadoun, T. (2005). Deformation and damage upon stretching of degradable polymers (PLA and PCL), *Polymer* (Guildf), 46(18), pp. 7370–85, Doi: 10.1016/J.POLYMER.2005.03.116.

[61] Li, Y., Volland, C., Kissel, T. (1998). Biodegradable brush-like graft polymers from poly(D,L-lactide) or poly(D,L-lactide-coglycolide) and charge-modified, hydrophilic dextrans as backbone – In-vitro degradation and controlled releases of hydrophilic macromolecules, *Polymer* (Guildf), 39(14), pp. 3087–97, Doi: 10.1016/S0032-3861(97)10048-9.

[62] Wang, Y.-C., Lin, M.-C., Wang, D.-M., Hsieh, H.-J. (2003). Fabrication of a novel porous PGA-chitosan hybrid matrix for tissue engineering, *Biomaterials*, 24(6), pp. 1047–57, Doi: 10.1016/S0142-9612(02)00434-9.

[63] Hsieh, C.-Y., Tsai, S.-P., Wang, D.-M., Chang, Y.-N., Hsieh, H.-J. (2005). Preparation of γ-PGA/chitosan composite tissue engineering matrices, *Biomaterials*, 26 (28), pp. 5617–23, Doi: 10.1016/j.biomaterials.2005.02.012.

[64] Zheng, Y.F., Gu, X.N., Witte, F. (2014). Biodegradable metals, *Mater. Sci. Eng. R Rep.*, 77, pp. 1–34, Doi: 10.1016/j.mser.2014.01.001.

[65] Aghion, E. (2018). Biodegradable metals, *Metals* (Basel), 8(10), p. 804, Doi: 10.3390/met8100804.

[66] Hermawan, H. (2012). *Biodegradable metals: State of the art*, Berlin, Heidelberg, Springer, pp. 13–22.

[67] Singh Raman, R.K., Jafari, S., Harandi, S.E. (2015). Corrosion fatigue fracture of magnesium alloys in bioimplant applications: A review, *Eng. Fract. Mech.*, 137, pp. 97–108, Doi: 10.1016/j.engfracmech.2014.08.009.

[68] Poinern, G.E.J., Brundavanam, S., Fawcett, D. (2012). Biomedical magnesium alloys: A review of material properties, surface modifications and potential as a biodegradable orthopaedic implant, *Am. J. Biomed. Eng.*, 2, pp. 218–40.

[69] Maguire, M.E., Cowan, J.A. (2002). Magnesium chemistry and biochemistry, *Bio-Metals*, 15, pp. 203–10.

[70] Touyz, R.M. (2004). Magnesium in clinical medicine, *Front. Biosci.*, 9, pp. 1278–93.

[71] Saris, N.E., Mervaala, E., Karppanen, H., Khawaja, J.A., Lewenstam, A. (2000). Magnesium. An update on physiological, clinical and analytical aspects, *Clin. Chim. Acta.*, 294(1–2), pp. 1–26.

[72] Gums, J.G. (2004). Magnesium in cardiovascular and other disorders., *Am. J. Health. Syst. Pharm.*, 61(15), pp. 1569–76.

[73] Walker, J., Shadanbaz, S., Woodfield, T.B.F., Staiger, M.P., Dias, G.J. (2014). Magnesium biomaterials for orthopedic application: A review from a biological perspective, *J. Biomed. Mater. Res. B. Appl. Biomater.*, 102(6), pp. 1316–31, Doi: 10.1002/jbm.b.33113.

[74] Liu, D., Yang, D., Li, X., Hu, S. (2018). Mechanical properties, corrosion resistance and biocompatibilities of degradable Mg-RE alloys: A review, *J. Mater. Res. Technol.*, Doi: 10.1016/J.JMRT.2018.08.003.

[75] Mccord, C.P., Prendergast, J.J., Meek, S.F., Harrold, G.C. (1942). Chemical gas gangrene from metallic magnesium, *Indust. Med.* [Beloit], 11(2), pp. 71–6.

[76] Wen, C., Mabuchi, M., Yamada, Y., Shimojima, K., Chino, Y., Asahina, T. (2001). Processing of biocompatible porous Ti and Mg, *Scr. Mater.*, 45(10), pp. 1147–53, Doi: 10.1016/S1359-6462(01)01132-0.

[77] Logan, H.L. (1958). Mechanism of Stress-Corrosion Cracking in the AZ31B Magnesium Alloy, *J. Res. Natl. Bur. Stand. (1934).*, 61(6), pp. 503–8.

[78] Bhuiyan, M.S., Mutoh, Y., Murai, T., Iwakami, S. (2008). Corrosion fatigue behavior of extruded magnesium alloy AZ61 under three different corrosive environments, *Int. J. Fatigue*, 30(10), pp. 1756–65, Doi: 10.1016/j.ijfatigue.2008.02.012.

2 Challenges and Common Strategies

2.1 INTRODUCTION

The low strength of as-cast magnesium is represented as one of the drawbacks that hampers its deployment as a material for biomedical devices. Although researchers were able to obtain properties comparable or even higher than those of human bones (see Section 2.2. for the strategies employed), magnesium is yet far from its application in the biomedical field. This is due to its high corrosion rate in the electrolytically physiologic environment. The issues arising from the high corrosion rate are manifold. First, an exceedingly high corrosion rate will lead to the desorption of the implant before the healing process is complete. Second, the continuous desorption of the material can lead to an untimely failure of the implant. Furthermore, H_2 gas evolution during the corrosion process may lead to the twofold problem of the embrittlement of the implant due to the hydrogen embrittlement phenomenon [1] and of hydrogen pockets that induce biocompatibility issues since they are harmful to the surrounding tissues. Further biocompatibility issues arise from the presence of corrosion cracks and hydrogen release, hampering an adequate settlement environment for cells to survive [2]. This is due to (1) the falling out of corrosion products, rendering it difficult for cells to attach to the surface [3], (2) the increase in the pH due to the hydrogen release, inducing death of cells [4] and (3) the toxic osmolarity as a consequence of very high concentration of Mg^{2+} ions, resulting from the corrosion process. There is a direct relation between corrosion and biological performances and thus, aiming to increase the biocompatibility, corrosion resistance must be improved. Regarding the link between corrosion and mechanical properties, the reduced strength induced by corrosion because of hydrogen embrittlement and the presence of notches, cracks and other stress concentrators induced by nonuniform corrosion modes are apparent; however, no other relation appears to be present at first sight. Corrosion and mechanical properties are however intrinsically connected, and we see more correlating effects at play upon a more detailed investigation. To clarify this statement, the strategies to improve the corrosion properties and those to improve the mechanical properties are reported, allowing a better understanding of the mentioned correlation.

2.2 CORROSION MITIGATION STRATEGIES

The presence of second phases and impurities is known to accelerate the corrosion of the magnesium matrix due to the occurrence of galvanic cells (see Appendix A). Seeking to improve the corrosion resistance, many researchers have thus studied the effects of impurities and of the alloying elements. In addition, the grain size has also been shown to influence the corrosion behavior of Mg alloys.

2.2.1 IMPURITIES REMOVAL

Because of poor molten metal handling processes and due to the natural composition of the raw magnesium, some undesirable impurities, such as iron (Fe), nickel (Ni) and copper (Cu), are incorporated into the material. The effect of impurities on the corrosion behavior of Mg-based materials has thus gained great interest and has been intensively investigated in recent years [5,6]. Song [7] compared the amount of evolved hydrogen in Hank's solution for commercially pure and high-pure magnesium, stating the purification to reduce the hydrogen evolution rate from 26 to 0.008 ml/cm^2/day, in accordance with Hofstetter et al. [8], who reported that an ultra-high-purity ZX50 Mg alloy shows greater corrosion resistance when compared to high-purity alloys (almost three times) and to standard purity alloy (over an order of magnitude). The main difference in these alloys is their impurity level, as reported in Table 2.1.

Again, Shi et al. [9] discovered better corrosion resistance of high-purity Mg and its alloys by comparing low- and high-purity Mg and Mg1Al alloys (Figure 2.1) in salt immersion test (SIT) and salt spray test (SST). Finally, Li et al. [10] reported that specimens made from 99.99 wt % pure Mg have no mass loss even after 180 days of immersion in simulated body fluid.

The corrosion rate is in fact reported to increase 10–100 times if the concentration of impurities rises beyond the tolerance limit determined by the maximum solid solubility [11]. This solid solubility limit is defined as the extent to which an element will dissolve in base materials (magnesium in this case) without forming a different phase [12]. Thus, a higher solid solubility limit leads to more homogenously dispersed elements, whereas a lower one increases the number of separate phases within the magnesium matrix. The tolerance limits are usually low, and the limit at which segregation in pure metal (Fe) or Mg-intermetallic phases (Ni, Cu) occurs can be as low as 35–50 ppm for Fe, 100–300 ppm for Cu and 20–50 ppm for Ni [13]. It is fundamental to keep these elements under their tolerance levels, or to moderate their

TABLE 2.1

Impurity content of ZX50 alloys studied in ref. [8]

Alloy	Fe (ppm)	Cu (ppm)	Ni (ppm)
Standard-purity ZX50	42	9	8
High-purity ZX50	31	8	7
Ultra-high-purity ZX50	0.5	0.09	0.05

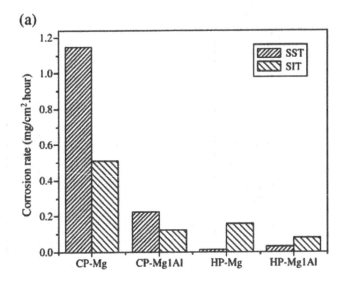

FIGURE 2.1 Comparison of the corrosion rates of the high-purity (HP) Mg with commercially pure (CP) Mg in the salt immersion test (3% NaCl SIT) and salt spray test (SST). Reprinted with permission from Elsevier [9].

activity utilizing alloying element (discussed later). However, the presence of alloying elements cross-influences their tolerance levels, as found by Hanawalt [14]. An element-specific impurity threshold level is thus not possible to be defined and it will be further discussed in Section 3.

2.2.2 ALLOYING

Some alloying elements are known to improve the corrosion resistance due to the "scavenger effect" and due to their passivation properties. Zinc and manganese, for example, help to overcome the harmful corrosive effect of iron and nickel impurities and other heavy-metal elements (scavenger effect). In addition, the formation of dense passive film layers on the surface can inhibit the chloride ion permeation and control matrix dissolution. However, if the solid solubility limit of the elements is exceeded, second phases would form, leading to the onset of a galvanic cell with the α-Mg matrix. In particular, most of the elements are characterized by a very low solubility (see Figure 2.2), and thus the insolubility of elements in magnesium represents a problematic scenario from a corrosion perspective.

2.2.3 GRAIN SIZE MODIFICATION

Grain refinement has been reported to improve the corrosion resistance of Mg alloys. Grain boundaries are characterized by higher imperfection and higher

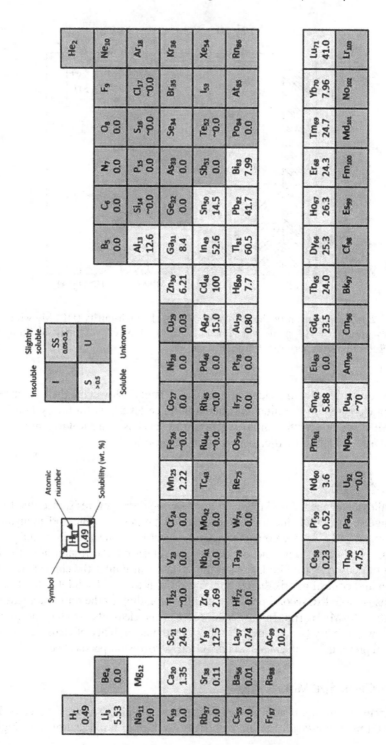

FIGURE 2.2 Maximum solubility of elements in magnesium (at any temperature) given in wt%.

internal energy compared to the Mg matrix; any corrosive attack takes place preferentially on grain boundaries. Segregation of alloying elements and second phases occurs on these boundaries, leading to an accelerated cathodic activity of the surrounding Mg matrix. This would normally favor coarse grains; however, such segregations are continuously distributed in Mg alloys, with finer grains leading to a more homogeneous corrosion behavior acting as a corrosion barrier due to their passivation behavior [13,15]. In addition, there are high compression stresses within the oxide layer due to geometrical mismatches with respect to the hexagonal Mg lattice, which cause cracks in the oxide, and introducing a large volume fraction of grain boundaries has been shown to compensate this effect, leading to improved corrosion responses [16]. Furthermore, the positive effect of small grain size on corrosion resistance has been attributed to the rapid formation of passivation oxide films at grain boundaries, which provides nucleation sites for passivation films, and the higher interfacial adherence of the passive film (MgO) at grain boundaries compared with the bulk [16,17]. Recently, Kim and Kim [18] observed the formation of many isolated MgO nanocrystals in the fine-grained Mg substrate below the MgO layer, which may have been formed as oxygen atoms diffused from the surface into the grain boundaries of the matrix. They proposed that the formation of the layer containing a mixture of MgO and Mg phases between the MgO layer and the Mg substrate is beneficial in decreasing the susceptibility of cracking of the MgO layer, or the interface between MgO and the Mg substrate by acting as a buffer layer that decreases the sharpness of the tensile stress gradient across the boundary between the MgO layer and the Mg substrate. Finally, fine grain sizes improve the corrosion-assisted cracking resistance, since they inhibit crack initiation and dislocation motion and lead to an increase in the number of barriers to crack propagation. High dislocation density has been found to decrease the electrochemical potential of the Mg matrix, thus increasing the anodic dissolution [19]. Ben Hamu et al. [20] reported the corrosion rate of AZ31 Mg alloys to be reduced by almost 17%, decreasing the grain size from 29–35 μm to 9–14 μm, in accordance with Liu et al. [21]who reported finer grain size to lead to lower weight losses (Figure 2.3).

Grain refinement can be mainly obtained in two different ways. First, leveraging on the fact that some alloying elements have a grain refinement effect. For example, the mechanical properties of as-cast Mg-2Zn have been reported to highly increase by addition of 0.2 wt% Zr. The tensile strength increases from 145.9 to 186.9 MPa, while the elongation to failure from 12.2% to 18% [22]. This will be discussed more in details in Section 3. The other, more effective, option is to act on the manufacturing process and on the cooling strategies [20,21,23–25]. In fact, inducing high plastic deformations within the material is known to lead to the nucleation and growth of new finer grains as a consequence of the dynamic recrystallization phenomenon.

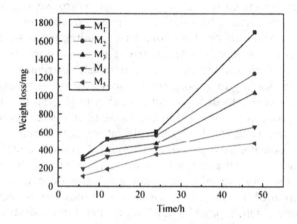

FIGURE 2.3 Weight loss of pure magnesium with different grain sizes immersed in the simulated body fluid. M1 data are characterized by the highest grain size, whereas M5 by the lowest. Reprinted with permission from Springer [21].

2.3 MECHANICAL PROPERTIES TUNING

As already mentioned, the poor mechanical properties of cast-pure Mg are a concern for their use in biomedical applications. In fact, many researchers studied the mechanical properties of a 99.99 wt% pure magnesium under tensile, bending and compressive loads [26–29], but the properties reported are far lower compared to those of human bones (Table 2.2).

To overcome this drawback of cast-pure Mg, researchers have started to investigate the improvement of the mechanical properties by means of different manufacturing processes, leveraging on the grain refinement effect. According to the famous Hall–Petch equation, the lower the grain size, the higher the yield strength. In addition, alloying elements are also known to improve the strength of the material due to the solid solution strengthening effect and precipitation hardening effect.

TABLE 2.2
Mechanical properties of cast 99.99 wt% magnesium [26] and of cortical bones [27]

Material	Yield stress (MPa)	Ultimate tensile strength (MPa)	Elong- ation to failure (%)	Compressive yield strength (MPa)	Ultimate compressive strength (MPa)
Pure magnesium	20.4	84.5	13	99	157.7
Cortical bones	110	175	—	—	200

2.3.1 GRAIN REFINEMENT

Usually, a metal is not made up of a single large crystal, but of many small crystals called grains, consisting in lattices each having a different orientation from the adjacent one. The formation of the grains begins during the solidification of the material: this last phase influences the characteristics of the grains in terms of size and quantity. The interfaces between the different crystalline lattices are called grain boundaries (Figure 2.4). The size and the orientation of the grains determine some mechanical properties of the material: the grain boundaries obstruct the plastic deformation, since the dislocations are strongly blocked in their movement near the interfaces. Their effect is greater the more the crystalline lattice orientation differs from one grain to another. Understandably, the material with fine grains has higher mechanical strength than the one with coarse grains; thus, the reduction of the grain size is an important hardening mechanism of the metal. Gu et al. [30] compared the yield strength and the ultimate tensile strength of as-cast and rolled 99.95 wt% pure Mg, reporting the rolling procedure to increase the yield strength and ultimate tensile strength of 460% and 97%, respectively, due to a reduction in the grain size of almost 90%.

The mathematical model representing this mechanism can be described with the Hall–Petch equation, which shows how, at temperatures lower than the recrystallization one, a fine-grained metal is stronger than a coarse-grained one:

$$\sigma_s = \sigma_0 + \frac{k}{\sqrt{d}} \tag{1}$$

where σ_s: yield strength.

σ_0, k: constants characteristic of the material.

d: average size of the crystalline grains.

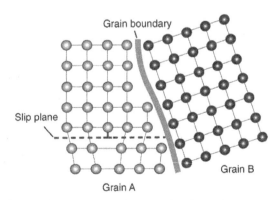

FIGURE 2.4 Effect of the grain boundary on the dislocation movement.

2.3.2 SOLID SOLUTION STRENGTHENING

Many alloys are solid solutions made by one or more metals dissolved in another metal:

- When the atoms of the alloying elements take the place of some atoms of the base metal in the crystalline lattice, there is a substitutional solid solution (see Figure 2.5).
- When these interpose themselves between the atoms of the lattice, it is called interstitial solid solution.

In both cases the lattice is distorted, delaying the movement of the dislocations and thus strengthening the material.

2.3.3 PRECIPITATION HARDENING

Precipitation hardening is one of the methods used to improve the mechanical properties of a metal. The precipitation hardening mechanism helps in increasing the difficulty of movement for dislocations due to appropriate distribution of particles at the grain boundaries and inside the grains. It is preferable to obtain a homogeneous distribution of particles, since a nonhomogeneous concentration of the precipitates leads to nonhomogeneous mechanical properties. The size of the particles is very important: small size particles give better mechanical properties. The hardening effect is due both to an increase in the difficulty of the dislocation motion and to an increase in the concentration of dislocations through a mechanism known as Orowan looping [31].

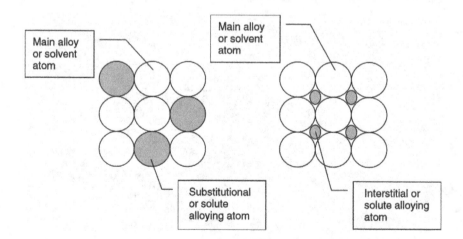

FIGURE 2.5   Schematic illustration of substitutional and interstitial alloying.

2.4 INTERPLAY AMONG MECHANICAL PROPERTIES, CORROSION RESISTANCE AND BIOCOMPATIBILITY

As reported at the beginning of this section, the corrosion behavior of the material influences its mechanical properties and, above all, biocompatibility. Mechanical properties and biocompatibility are not directly interlinked. For example, a material with higher stiffness will not necessarily change its biocompatibility, given that its composition stays the same. Moreover, as it can be understood from Section 2.3, the strategy adopted to overcome one of the main drawbacks of as-cast magnesium, that is, either low mechanical properties or low corrosion resistance, has direct impacts on the other. Grain refinement has a positive effect on both, increasing the mechanical properties and the corrosion resistance. Also, adding alloying elements is an effective way to improve the strength of the material, but it may lead to a decrease in the corrosion resistance if the solubility limit is exceeded. In Chapter 3, the main alloys studied for biomedical applications are reviewed in terms of their mechanical properties, corrosion resistance and biocompatibility, and in Chapter 4, a summary of their performance with respect to the challenges is reported.

REFERENCES

[1] Kappes, M., Iannuzzi, M., Carranza, R.M. (2013). Hydrogen embrittlement of magnesium and magnesium alloys: A review, *J. Electrochem. Soc.*, 160(4), pp. c168–78, Doi: 10.1149/2.023304jes.

[2] Kim, J., Mousa, H.M., Park, C.H., Kim, C.S. (2017). Enhanced corrosion resistance and biocompatibility of AZ31 Mg alloy using PCL/ZnO NPs via electrospinning, *Appl. Surf. Sci.*, 396, pp. 249–58, Doi: 10.1016/J.APSUSC.2016.10.092.

[3] Razavi, M., Fathi, M., Savabi, O., Vashaee, D., Tayebi, L. (2014). In vitro study of nanostructured diopside coating on Mg alloy orthopedic implants, *Mater. Sci. Eng. C*, 41, pp. 168–77, Doi: 10.1016/j.msec.2014.04.039.

[4] Razavi, M., Fathi, M., Savabi, O., Vashaee, D., Tayebi, L. (2015). *In vivo* study of nanostructured akermanite/PEO coating on biodegradable magnesium alloy for biomedical applications, *J. Biomed. Mater. Res. Part A*, 103(5), pp. 1798–808, Doi: 10.1002/jbm.a.35324.

[5] Murray, R.W., Hillis, J.E. (1990). *Magnesium finishing: Chemical treatment and coating practices.* SAE Technical Papers, SAE International.

[6] Hillis, J.E. (1983). The Effects of Heavy Metal Contamination on Magnesium Corrosion Performance, *SAE Trans.*, pp. 553–9, Doi: 10.2307/44668089.

[7] Song, G. (2007). Control of biodegradation of biocompatable magnesium alloys, *Corros. Sci.*, 49(4), pp. 1696–701, Doi: 10.1016/J.CORSCI.2007.01.001.

[8] Hofstetter, J., Becker, M., Martinelli, E., Weinberg, A.M., Mingler, B., Kilian, H., Pogatscher, S., Uggowitzer, P.J., Löffler, J.F. (2014). High-strength low-alloy (HSLA) Mg–Zn–Ca alloys with excellent biodegradation performance, *JOM*, 66 (4), pp. 566–72, Doi: 10.1007/s11837-014-0875-5.

[9] Shi, Z., Song, G., Atrens, A. (2006). Corrosion resistance of anodised single-phase Mg alloys, *Surf. Coatings Technol.*, 201(1), pp. 492–503, Doi: 10.1016/j.surfcoat.2005.11.081.

[10] Li, L., Gao, J., Wang, Y. (2004). Evaluation of cyto-toxicity and corrosion behavior of alkali-heat-treated magnesium in simulated body fluid, *Surf. Coatings Technol.*, 185(1), pp. 92–8, Doi: 10.1016/j.surfcoat.2004.01.004.

[11] Hillis, J., Murray, R. (1987). Finishing alternatives for high purity magnesium alloys, SDCE 14th International Die Casting Congress and Exposition, Toronto.

[12] Persaud-Sharma, D., McGoron, A. (2012). Biodegradable magnesium alloys: A review of material development and applications, *J. Biomim. Biomater. Tissue Eng.*, 12(2011), pp. 25–39, Doi: 10.4028/www.scientific.net/JBBTE.12.25.

[13] Witte, F., Hort, N., Vogt, C., Cohen, S., Kainer, K.U., Willumeit, R., Feyerabend, F. (2008). Degradable biomaterials based on magnesium corrosion, *Curr. Opin. Solid State Mater. Sci.*, 12(5), pp. 63–72, Doi: 10.1016/j.cossms.2009.04.001.

[14] Hanawalt, J.D., Nelson C.E., Peloubet, J.A. (1942). Corrosion studies of magnesium and its alloys, *Trans AIME*, 47, pp. 273–99.

[15] Singh Raman, R.K., Jafari, S., Harandi, S.E. (2015). Corrosion fatigue fracture of magnesium alloys in bioimplant applications: A review, *Eng. Fract. Mech.*, 137, pp. 97–108, Doi: 10.1016/j.engfracmech.2014.08.009.

[16] Birbilis, N., Ralston, K.D., Virtanen, S., Fraser, H.L., Davies, C.H.J. (2010). Grain character influences on corrosion of ECAPed pure magnesium, *Corros. Eng. Sci. Technol.*, 45(3), pp. 224–30, Doi: 10.1179/147842209X12559428167805.

[17] Ralston, K.D., Birbilis, N., Davies, C.H.J. (2010). Revealing the relationship between grain size and corrosion rate of metals, *Scr. Mater.*, 63(12), pp. 1201–4, Doi: 10.1016/J.SCRIPTAMAT.2010.08.035.

[18] Kim, H.S., Kim, W.J. (2013). Enhanced corrosion resistance of ultrafine-grained AZ61 alloy containing very fine particles of Mg17Al12 phase, *Corros. Sci.*, 75, pp. 228–38, Doi: 10.1016/J.CORSCI.2013.05.032.

[19] Ben-Hamu, G., Eliezer, A., Gutman, E.M. (2006). RETRACTED: Electrochemical behavior of magnesium alloys strained in buffer solutions, *Electrochim. Acta*, 52(1), pp. 304–13, Doi: 10.1016/J.ELECTACTA.2006.05.009.

[20] Hamu, G.B., Eliezer, D., Wagner, L. (2009). The relation between severe plastic deformation microstructure and corrosion behavior of AZ31 magnesium alloy, *J. Alloys Compd.*, 468(1–2), pp. 222–9, Doi: 10.1016/J.JALLCOM.2008.01.084.

[21] Liu, Y., Liu, D., You, C., Chen, M. (2015). Effects of grain size on the corrosion resistance of pure magnesium by cooling rate-controlled solidification, *Front. Mater. Sci.*, 9(3), pp. 247–53, Doi: 10.1007/s11706-015-0299-3.

[22] Zheng, Y.F., Gu, X.N., Witte, F. (2014). Biodegradable metals, *Mater. Sci. Eng. R Rep.*, 77, pp. 1–34, Doi: 10.1016/j.mser.2014.01.001.

[23] Argade, G.R., Panigrahi, S.K., Mishra, R.S. (2012). Effects of grain size on the corrosion resistance of wrought magnesium alloys containing neodymium, *Corros. Sci.*, 58, pp. 145–51, Doi: 10.1016/J.CORSCI.2012.01.021.

[24] Zhao, Y.-C., Huang, G.-S., Wang, G.-G., Han, T.-Z., Pan, F.-S. (n.d.). Influence of grain orientation on the corrosion behavior of rolled AZ31 magnesium alloy, *Acta Metall. Sin. (English Lett.)*, 28, Doi: 10.1007/s40195-015-0337-2.

[25] Bertolini, R., Bruschi, S., Ghiotti, A., Pezzato, L., Dabalà, M. (2017). The effect of cooling strategies and machining feed rate on the corrosion behavior and wettability of AZ31 alloy for biomedical applications, *Procedia CIRP*, 65, pp. 7–12, Doi: 10.1016/J.PROCIR.2017.03.168.

[26] Zhou, Y.-L., Li, Y., Luo, D.-M., Ding, Y., Hodgson, P. (2015). Microstructures, mechanical and corrosion properties and biocompatibility of as extruded Mg–Mn–Zn–Nd alloys for biomedical applications, *Mater. Sci. Eng. C*, 49, pp. 93–100, Doi: 10.1016/j.msec.2014.12.057.

[27] Trivedi, P., Nune, K.C., Misra, R.D.K. (2016). Degradation behaviour of magnesium-rare earth biomedical alloys, *Mater. Technol.*, 31(12), pp. 726–31, Doi: 10.1080/10667857.2016.1213550.

[28] Pan, Y., He, S., Wang, D., Huang, D., Zheng, T., Wang, S., Dong, P., Chen, C. (2015). In vitro degradation and electrochemical corrosion evaluations of microarc

oxidized pure Mg, Mg–Ca and Mg–Ca–Zn alloys for biomedical applications, *Mater. Sci. Eng. C*, 47, pp. 85–96, Doi: 10.1016/J.MSEC.2014.11.048.

[29] Wan, Y., Xiong, G., Luo, H., He, F., Huang, Y., Zhou, X. (2008). Preparation and characterization of a new biomedical magnesium–calcium alloy, *Mater. Des.*, 29 (10), pp. 2034–7, Doi: 10.1016/J.MATDES.2008.04.017.

[30] Gu, X., Zheng, Y., Cheng, Y., Zhong, S., Xi, T. (2009). In vitro corrosion and bio-compatibility of binary magnesium alloys, *Biomaterials*, 30(4), pp. 484–98, Doi: 10.1016/J.BIOMATERIALS.2008.10.021.

[31] Sun, F., Gu, Y.F., Yan, J.B., Zhong, Z.H., Yuyama, M. (2016). Phenomenological and microstructural analysis of intermediate temperatures creep in a Ni–Fe-based alloy for advanced ultra-supercritical fossil power plants, *Acta Mater.*, 102, pp. 70–8, Doi: 10.1016/J.ACTAMAT.2015.09.006.

3 Synopsis of Properties of Biocompatible Mg and Its Alloys

3.1 INTRODUCTION

The development of Mg materials in the field of biodegradable applications has been primarily focused on improving the corrosion resistance. The development of high-pure magnesium with a level of impurities below a critical threshold, along with the addition of alloying elements that either decrease the amount of impurities or refine the grain size, has been considered. Further, thermo-mechanical processes have been used to decrease the corrosion rate. In this chapter, the main classes of biocompatible Mg alloys are listed and discussed, reporting on their mechanical properties, corrosion behavior and biological compatibility, when available. For the sake of conciseness, this chapter only includes data on Mg alloys that are part of the discussion. For additional information on the state of the art, the reader is referred to the online version of the tables on www.routledge.com/9780367429454.

3.2 HIGH-PURE MAGNESIUM

The harmful effects of second phases and impurities on corrosion resistance have been reported in the previous section. As mentioned, one way to improve corrosion performance is to minimize the impurity level, and researchers have thus focused on obtaining high-pure magnesium reaching a 99.99% ultra-high-pure magnesium so far [1]. In this case, the amount of impurities is far below their tolerance level (Fe content of about 2.2 ppm, while Ni and Cu lower than 1 ppm).

3.2.1 HIGH-PURE MAGNESIUM: MECHANICAL PROPERTIES

As mentioned in Chapter 2, the mechanical properties of cast-pure and high-pure Mg are far lower than those of cortical bones (Table 2.2, Chapter 2). To overcome this drawback of cast-pure Mg, researchers have started to investigate the improvement of the mechanical properties by means of different manufacturing processes, leveraging on the grain refinement effect. Jeong and Kim [2] then compared cast and extruded Mg, reporting an increment from 27.8 to 100.6 MPa and from 89.7 to 179.5 MPa for yield strength and ultimate tensile strength (UTS), respectively, while the elongation to failure was reduced from 22.5% to 13.7%. Other researchers have studied the mechanical properties of extruded pure Mg [3–5], obtaining good agreement with those reported above. Moreover, some of them studied new manufacturing process to enhance both the mechanical and corrosion properties of pure Mg via grain refinement, obtaining discrepancies in the results. Seong and Kim found a reduction in the yield strength to that reported above in a 99.9 wt.% extruded pure Mg when employing a high-ratio differential speed rolling (HRDSR) and a post-rolling annealing after the extrusion process [5]. In contrast, manufacturing a 99.5+ wt.% Mg by means of hydrostatic extrusion leads to a yield strength of 178 MPa and a UTS of 228 MPa, respectively [6]. Other manufacturing procedures have then been developed, from the equal channel angular pressing (ECAP) [7,8] to Mg powder sintering procedures [9,10]. However, although the mechanical properties of pure Mg equals those of cortical bones and in some cases also slightly exceed it (Table 3.1), the increment is still too low since the mechanical integrity cannot be guaranteed to be maintained for the whole healing time. Saha et al. [11] found a reduction of almost 46% in the compressive strength of a 99.97 wt.% pure Mg after soaking in Hank's balanced salt solution (HBSS) for 16 days. Degradation of the mechanical properties has been studied both *in vivo* and *in vitro* [12].

TABLE 3.1
Summary of high-pure Mg mechanical properties.*

Manufacturing process	Yield strength (MPa)	UTS (MPa)	Elongation to failure (%)	Bending strength (MPa)	Compressive yield strength (MPa)	Ultimate compressive strength (MPa)	Ref.
Cast	27.8	89.7	22.5	—	—	—	[2]
Extruded	100.6	179.9	13.7	—	—	—	[2]
Extruded	39.7	249.3	—	—	—	—	[3]
Extruded	107.2	197.1	10.3	—	—	—	[4]
Extruded	100.6	—	—	—	—	—	[5]
HRDSR + annealed	27.1	—	—	—	—	—	[5]
Hydrostatic extruded	178	228	19	—	—	—	[6]
Hydrostatic extruded + back pressure	152	230	11	—	—	—	[6]
ECAP	56.8	162.6	14.8	—	—	—	[7,8]
Microwave-assisted sintering	—	132.7	—	—	—	—	[9]
Sintering + extrusion	—	—	—	—	224	340	[10]
Cast	—	—	—	—	—	227	[11]

* All the Mg samples are cast, unless otherwise specified. Additional information can be found on www.routledge.com/9780367429454

3.2.2 HIGH-PURE MAGNESIUM: CORROSION RESISTANCE

Reducing the corrosion rate of magnesium is fundamental for its application as a biocompatible material. High corrosion rates lead to acute inflammation related to hydrogen gas release accompanying dissolution. Moreover, changes in local pH due to magnesium dissolution can compromise the viability of the cells [13]. Based on a study on the implanted femur rods in guinea pigs [14], Song [15] determined 0.01 ml/cm^2/day as the tolerance level for hydrogen evolution. Liu et al. [16] reported that the purification of magnesium significantly hampers its corrosion. They evaluated the volume of evolved H_2 for both low-purity (LP) magnesium with an Fe content (280 ppm) above its tolerance limit and high-purity (HP) magnesium characterized by an Fe content of 45 ppm, both soaked in 3% NaCl for 33 h. The results are gathered in Figure 3.1.

The results are in qualitative accordance with ref. [1], where the focus was on the ratio of two specific impurity elements (Fe/Mn) rather than the absolute value of each individually. Several studies in literature deal with the improvement of the corrosion resistance of Mg alloys by adding Mn as alloying element since it increases the Fe tolerance level [17–19]. Lee et al. [20] studied the effects of different impurity contents on magnesium, and reported that specimens with higher Fe content but lower Fe/Mn ratio yield lower H_2 evolution. In addition, Abidin et al. [21] demonstrated that Mg alloys including ZE41 and AZ91 corrode faster than high-purity Mg in Hank's solution at 37°C. This is in agreement with the *in vitro* tests of Li et al. [22] and with the *in vivo* studies of Zhang et al. [23]. The effect of different thermo-mechanical treatments has been studied well. Gu et al. [24] reported that hot rolling reduced the corrosion current density of 99.95% pure magnesium due to grain refinement similarly as in ref. [11], where ECAP,

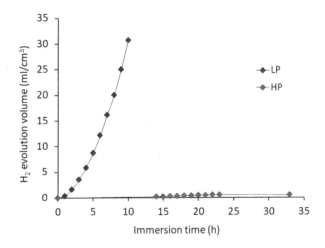

FIGURE 3.1 Hydrogen evolution volume for LP and HP magnesium. Modified from ref. [16].

rolling and extrusion lead to related effects. However, annealing increases corrosion current density as explained by Liu et al. [16] with respect to the Fe tolerance level. As a result, cast Mg has a tolerance level of 180 ppm due to Fe's confinement in supersaturated solid solutions inside the Mg lattice. After a heat treatment, however, the Fe tolerance drops to almost 10 ppm due to Fe's precipitation. This suggests that different environments lead to different corrosion behavior as revealed through a change in polarization characteristics. Walker et al. [25,26] compared the weight loss of commercial and high-pure magnesium in three different environments for a maximum soaking time of four weeks, reporting differences in Mg weight loss up to 130%. Corrosion data in terms of electrochemical polarization tests and corrosion rate are gathered in Tables 3.2 and 3.3, respectively. In Table 3.3, details on the procedures used for the corrosion rate determination are provided in the "procedure" column, where "H" stands for the derivation of the corrosion rate from the evolved hydrogen, whereas "WL" means that the corrosion rate has been obtained by means of Equation A.1.

Nguyen et al. [28] evaluated the effects of surface roughness on the corrosion rate after 6 h of soaking in HBSS at 37°C. Surface roughness favors cell adhesion and growth [29–32] but increases the corrosion rate (Figure 3.2), which is in agreement with ref. [33].

3.2.3 High-Pure Magnesium: Biocompatibility

Li et al. [34] assessed the effects of the degradation products on cell proliferation and osteogenic differentiation of human mesenchymal stem cells, reporting that magnesium has the ability to favor the formation of new bone, which is in agreement with a variety of other studies [3,11,35,36]. Xue et al. implanted 99.9% pure Mg in the subcutaneous tissue of mice and after two

---

TABLE 3.2

Summary of the DC polarization curve results for different purity levels of magnesium

Purity (%)	Fe/Mn ratio	Corrosive environment	E_{corr} (V)	i_{corr} (μA/cm^2)	Ref.
99.9 (HRDSR + annealed)	0.08	Hank's, 37°C	−1.5	13.47	[5]
99.75	—	SBF, 37°C	−1.78	251	[9]
99.99	0.35	Hank's, 37°C	−1.63	4	[21]
99.99	—	Hank's, 37°C	−1.73	200	[7]
99.99 (ECAP)	—	Hank's, 37°C	−1.56	40	[7]
99.95 (extruded)	—	SBF, 37°C	−1.8	51.34	[4]
CP	—	SBF, 37°C	−1.90	363.21	[23]

* All the Mg samples are cast, unless otherwise specified. Additional information can be found on www.routledge.com/9780367429454.

TABLE 3.3
Summary of the corrosion rates for different purity levels of magnesium.*

Purity (%)	Fe/Mn ratio	Corrosive environment	Immersion time	Corrosion rate (mm/year)	Procedure	Ref.
99.99	—	*In vivo* (Lewis rates)	7 days	0.390	WL	[25]
99.99	—	*In vivo* (Lewis rates)	14 days	0.390	WL	[25]
99.99	—	*In vivo* (Lewis rates)	21 days	0.221	WL	[25]
99.99	—	EBSS	7 days	0.572	WL	[25]
99.99	—	EBSS	14 days	0.468	WL	[25]
99.99	—	EBSS	21 days	0.382	WL	[25]
99.99	—	MEM	7 days	0.728	WL	[25]
99.99	—	MEM	14 days	0.676	WL	[25]
99.99	—	MEM	21 days	0.659	WL	[25]
99.99	—	MEMp	7 days	2.185	WL	[25]
99.99	—	MEMp	14 days	1.483	WL	[25]
99.99	—	MEMp	21 days	1.370	WL	[25]
HP	6	3% NaCl	225 h	1.8	H	[16]
HP (heat treated 1 day at 550°C)	6	3% NaCl	225 h	1.8	H	[16]
HP (heat treated 2 days at 550°C)	6	3% NaCl	225 h	1.8	H	[16]
99.98	—	HBSS, 37°C	8 days	0.32	H	[3]
99.9 (HRDSR + annealing)	0.08	Hank's, 37°C	1 day	0.7	WL	[5]
99.9 (HRDSR + annealing)	0.08	Hank's, 37°C	7 days	0.57	WL	[5]
CP	—	SBF, 37°C	14 days	6.8	WL	[23]
99.75	—	SBF, 37°C	10 days	3.82	H	[9]
99.99	0.35	Hank's, 37°C	14 days	0.08	H	[21]
99.99	0.35	Hank's, 37°C	12 days	0.05	H	[21]
99.97	—	HBSS, 37.4°C	16 days	6.09	WL	[11]
99.97	—	HBSS, 37.4°C	32 days	4.82	WL	[11]
99.99	—	SBF, 37°C	8 days	5.86	H	[27]

* All samples are to be considered cast, unless otherwise specified. Additional information can be found on www.routledge.com/9780367429454.

months of implantation, no toxicity issues were noticeable. However, from *in vitro* studies, the cytocompatibility of pure Mg is still debated, since discordant results are reported in literature. Table 3.4 shows that not all cell types have viabilities higher than 70% (the value below which a material is

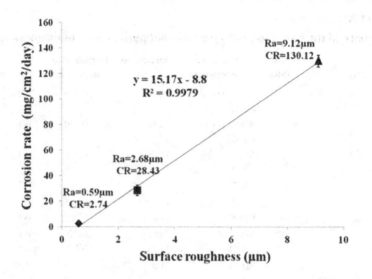

FIGURE 3.2 Corrosion rate as a function of surface roughness. Reprinted with permission from Wiley [28].

considered toxic). As a matter of fact, pure Mg is not suitable to be used in stent application since the hemolysis is far below 95%, which is the threshold for the onset of cytotoxic effects. In conclusion, varying cytotoxic responses and low mechanical properties led researchers to redirect effects to alloying for achieving biological compatibility.

3.3   ALUMINUM-BASED ALLOYS

Al is the most common addition to magnesium, it is relatively cheap, light, soluble and improves strength considerably (i.e., from 170 to 250 MPa in AZ91) [37] due to the mechanisms of solid solution and precipitation strengthening. In addition, it is passivating and improves corrosion resistance because of the formation of a passivating Al_2O_3 layer. However, increasing Al concentration does not correspond to a decrease in corrosion rate. Song et al. [38] tested different Mg–Al alloys in chloride solution and discovered that AZ91 alloys have higher corrosion rates than AZ21 because of the presence of second phases ($Mg_{17}Al_{12}$). Gusieva et al. [39] reported that alloys of higher Al content than AZ31 have second phases, implying that Al amounts above 3 wt.% lower the corrosion resistance. Some authors contradicted this hypothesis stating that an increase in Al steadily rises corrosion resistance [40,41]. Lunder et al. [40] suggested that anodic dissolution is further decreased with Al above 10 wt.%. Winzer et al. [42] resumed these studies observing two influences of $Mg_{17}Al_{12}$ phases on corrosion. They act (1) as a barrier and (2) as galvanic cathodes depending on the amount of second phases and on their distribution. $Mg_{17}Al_{12}$ accelerates corrosion at low volume fractions, but when

TABLE 3.4

Summary of the biocompatibility of pure Mg

Purity (%)	Fe/Mn ratio	Test environment	Cell viability			Platelet adhesion	Hem-olysis (%)	Ref.
			Cell type, procedure	Time of culture	Result (%)			
99.95	—	DMEM, 37°C	L-929, IC	2	64.70	—	—	[24]
99.95	—	DMEM, 37°C	L-929, IC	4	65.74	—	—	[24]
99.95	—	DMEM, 37°C	NIH3T3, IC	2	91.32	—	—	[24]
99.95	—	DMEM, 37°C	NIH3T3, IC	4	89.81	—	—	[24]
99.95	—	DMEM, 37°C	NIH3T3, IC	7	90.94	—	—	[24]
99.95	—	DMEM, 37°C	MC3T3-E1, IC	2	61.66	—	—	[24]
99.95	—	DMEM, 37°C	MC3T3-E1, IC	4	95.36	—	—	[24]
99.95	—	DMEM, 37°C	MC3T3-E1, IC	7	87.70	—	—	[24]
99.95	—	DMEM, 37°C	ECV304, IC	2	85.61	—	—	[24]
99.95	—	DMEM, 37°C	ECV304, IC	4	79.96	—	—	[24]
99.95	—	DMEM, 37°C	ECV304, IC	7	76.69	—	—	[24]
99.95	—	DMEM, 37°C	VSMC, IC	2	95.56	—	—	[24]
99.95	—	DMEM, 37°C	VSMC, IC	4	96.27	—	—	[24]
99.95	—	DMEM, 37°C	VSMC, IC	7	94.92	—	—	[24]
99.95	—	Human blood, 37°C	—	—	—	—	57	[24]
99.95 (rolled)	—	Human blood, 37°C	—	—	—	—	25	[24]
99.95	—	PRP, 37°C	—	—	—	23,608 per mm^2	—	[24]
99.95 (rolled)	—	PRP, 37°C	—	—	—	21,686 per mm^2	—	[24]
99.98	—	DMEM, 37°C	MG63, IC	1	96.39	—	—	[3]
99.98	—	DMEM, 37°C	MG63, IC	3	90.49	—	—	[3]
99.98	—	DMEM, 37°C	MG63, IC	7	115.41	—	—	[3]

* All samples are cast, unless otherwise specified. Additional information can be found on www.routledge.com/9780367429454.

forming an interconnected network at high fractions, it reduces corrosion acting as a barrier through the passivating properties of Al (Figure 3.3).

However, the extensive knowledge obtained with Al alloys is not directly applicable to biomedical implants. Long-term effects of exposure to Al reveals that Al is toxic, affecting the reproductive ability [43], inducing dementia [44] and leading to Alzheimer's disease [45,46]. Despite their limited suitability for biomedical applications, Al-based Mg alloys are the most studied, and the findings related to these alloys are significant and highly relevant to the field. Al-based alloys can be divided into three main series: AZ alloys, AM alloys and Mg–Al–rare earth (RE) alloys, which are discussed separately in the following sections.

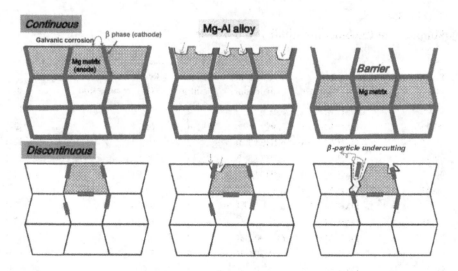

FIGURE 3.3 Schematic illustration of the role of $Mg_{17}Al_{12}$ phases in the Mg matrix when distribution is continuous or discontinuous, respectively. Reprinted with permission from Elsevier [2].

3.3.1 AZ ALLOYS

AZ alloys are characterized by Al and Zn as main alloying elements. The most studied are AZ21, AZ61, AZ80, AZ31 and AZ91. In the following sections, the mechanical properties, corrosion behavior and biocompatibility of AZ alloys are extensively reviewed.

3.3.1.1 AZ Alloys: Mechanical Properties

In the Table 3.5, the mechanical properties such as tensile and compressive yield strength, tensile and compressive strength and elongation to failure of different AZ alloys are gathered. Different manufacturing processes lead to different mechanical properties. In particular, the properties of as-cast AZ alloys can be improved either by refining the grain size or by continuous precipitation of $Mg_{17}Al_{12}$. Certain studies deal with the capability of RE elements to enhance the mechanical properties of Mg–Al–Zn alloys. For example, adding 0.5 wt.% of Ce to as-cast AZ91 increases the yield strength of about 13%, whereas it has a negligible effect on the tensile strength [47]. When there is too much Ce present (above 1.5% in [48]), a coarsening of Al_4Ce particles negatively affects the mechanical properties of the alloy. The mechanical properties of AZ alloys are reviewed in Table 3.5, the addition of RE elements, where applicable, is specified.

Increasing the Al content does not always lead to an increment in the mechanical properties. About 50% increase is obtained both in ultimate tensile strength (UTS) (from 135 to ~203 MPa) and yield strength (from 65 to ~96

TABLE 3.5
Summary of AZ alloys mechanical properties

Alloy	Add-itional element	Yield strength (MPa)	UTS (MPa)	Elong-ation to failure (%)	Compres-sive yield strength (MPa)	Ultimate compressive strength (MPa)	Ref.
AZ31 (hydrostatically extruded)	—	343	396	10	—	—	[6]
AZ31 (extruded + back pressure)	—	384	427	10	—	—	[6]
AZ61 (hydrostatically extruded)	—	418	445	9	—	—	[6]
AZ61 (extruded + back pressure)	—	443	487	8	—	—	[6]
AZ91 (hydrostatically extruded)	—	424	455	5	—	—	[6]
AZ91 (extruded + back pressure)	—	443	498	7	—	—	[6]
AZ01	—	64	140	2.5	—	—	[49]
AZ21	—	95	210	3.5	—	—	[49]
AZ41	—	97	209	3.1	—	—	[49]
AZ61	—	87	208	2.9	—	—	[49]
AZ91	—	83	169	2.4	—	—	[49]
AZ80	—	96	203	—	—	—	[50]
AZ80	0.5% Bi	99	210	—	—	—	[50]
AZ80	1% Bi	100	202	—	—	—	[50]
AZ80	2% Bi	101	160	—	—	—	[50]
AZ80 (aged)	—	122	224	—	—	—	[50]
AZ80 (aged)	0.5% Bi	132	225	—	—	—	[50]
AZ80 (aged)	1% Bi	118	216	—	—	—	[50]
AZ80 (aged)	2% Bi	113	210	—	—	—	[50]
AZ61 (extruded)	—	171	296	11.16	—	—	[48]
AZ61 (extruded)	0.5% Ce	172	299	12.42	—	—	[48]
AZ61 (extruded)	1% Ce	180	302	14.22	—	—	[48]
AZ61 (extruded)	1.5% Ce	165	288	14.15	—	—	[48]
AZ61 (annealed)	—	162	291	12.99	—	—	[48]
AZ61 (annealed)	0.5% Ce	165	292	13.08	—	—	[48]
AZ61 (annealed)	1% Ce	167	301	16.48	—	—	[48]
AZ61 (annealed)	1.5% Ce	166	290	11.994	—	—	[48]
AZ91	—	94	168	3.03	—	—	[47]
AZ91	1% RE	106	166	3.73	—	—	[47]
AZ91	2% RE	105	169	3.14	—	—	[47]
AZ91	3% RE	107	167	2.76	—	—	[47]

* All the AZ samples are cast, unless otherwise specified. Additional information can be found on
  www.routledge.com/9780367429454.

MPa) by addition of 2% Al to an AZ01 alloy. Further Al additions (up to 6.5%) did not cause any significant change in both UTS and yield strength. On the contrary, UTS and yield strength decrease in an AZ91 alloy. On the other hand, elongation continuously decreases for above 2% Al content [49]. The reduction in both elongation and strength has been reported to result from net-like intermetallics ($Mg_{17}Al_{12}$) along the α-Mg grain boundary. According to Candan et al. [51], the presence of β intermetallics yields a continuous crack path of brittle phases along the α-Mg grain boundary, leading to a low elongation and thus to a concomitant low strength. The effect of RE or other alloying elements are manifold. For example, adding 0.5% Bi to an AZ80 alloy increases its tensile and yield strength by almost 10% due to the dispersion strengthening of Mg_3Bi_2. However, a further increment in Bi content determines a decrease in the mechanical properties because coarse flaky Mg_3Bi_2 phases are prone to split the α-Mg matrix [50].

3.3.1.2   AZ Alloys: Corrosion Resistance

The passivation of Al leads to the formation of an Al_2O_3 layer on the surface of the material, which, unlike $Mg(OH)_2$, is insoluble in chloride-containing solutions. In addition, Zn has been reported to reduce the amount of hydrogen gas, since Zn^{2+} ions compete with the Mg^{2+} ions in solution for binding with free OH^- anions, forming $Zn(OH)_2$ that decreases the amount of H_2 gas [52,53]. However, the effect of impurities and second phases ($Mg_{17}Al_{12}$) affect the corrosion resistance of AZ alloys. The latter has already been described in Section 3.3, whereas the former has widely been investigated in literature, confirming the harmful effects of Fe impurities. In particular, the Fe/Mn ratio rather than the amount of Fe determines the corrosive behavior of Al-based alloys. A weight ratio of 0.032 of Fe/Mn is widely defined to be the threshold, above which the corrosion rate highly increases [17–19]. For improving the corrosion resistance of AZ alloys, researchers put efforts in controlling the Fe/Mn ratio; some studies deal with the addition of RE materials [54]. Luo et al. [55] studied the influence of yttrium on the corrosion resistance of AZ91 alloy, and reported a concentration of 0.3 wt.% decreases the corrosion rate. Similar effects were observed when adding 0.5 wt.% Sm to aged AZ92 alloy [56]. The corrosion characteristics of AZ alloys are gathered in Tables 3.6 (polarization curves) and 3.7 (corrosion rate), reporting, when available, the Fe/Mn ratio and the addition of other alloying elements.

Apart from the influence of the Fe/Mn ratio on the corrosion resistance, the manufacturing processes have a broad influence as well since they induce microstructural modifications [57–65]. Finer grain size has been widely reported to reduce corrosion. Bukovinovà and Hadzima [66] studied the corrosion behavior of die-cast AZ31 and extruded AZ31 alloys in Hank's solution at 37°C, respectively. They reported the latter to have a higher corrosion resistance as well as a uniform corrosion mode. Thus, motivated by the corrosion-reducing effects of grain refinement, researchers carried out *in vitro* corrosion experiments on AZ alloys manufactured by grain refining processes such as ECAP; however, the results obtained are inconsistent. Wang et al. [67], Gu

TABLE 3.6

Summary of the DC polarization curve results for AZ series Mg alloys.*

Alloy	Additional elements	Fe/Mn ratio	Corrosive environment	E_{corr} (V)	i_{corr} ($\mu A/cm^2$)	Ref.
AZ31 (rolled)	—	0.025	SBF, 37°C	−1.70	61.70	[57]
AZ31 (heat treated + quenched)	—	—	SBF, 37°C	−1.68	230.4	[58]
AZ91 (heat treated + quenched)	—	—	SBF, 37°C	−1.71	150.1	[58]
AZ31 (extruded)	—	—	HBSS, 37°C	−1.48	74.2	[59]
AZ31 (extruded)	—	—	Hank's, 37°C	−1.40**	81.5**	[60]
AZ31 (extruded)	—	—	DMEM, 37°C	−1.39**	15.1**	[60]
AZ31 (extruded)	—	—	DMEM + FBS, 37°C	−1.35**	4.7**	[60]
AZ91 (extruded)	—	—	Hank's, 37°C	−1.34**	80.8**	[60]
AZ91 (extruded)	—	—	DMEM, 37°C	−1.32**	21.1**	[60]
AZ91 (extruded)	—	—	DMEM + FBS, 37°C	−1.10**	19.1**	[60]
AZ31	—	0.05	SBF 5×, 37°C	−1.59	24,700	[61]
AZ31 (ECAP)	—	0.05	SBF 5×, 37°C	−1.52	6220	[61]
AZ31 (hot rolled)	—	—	DMEM, 37°C	−1.54	27.5	[62]
AZ31 (extruded)	—	—	Hank's, 37°C	−1.6	25.1	[63]

* All the AZ samples are cast, unless otherwise specified. Additional information can be found on www.routledge.com/9780367429454. ** Measured after 7 days immersion,

et al. [68] and Sunil et al. [61] reported ECAP to lower the degradation rates and the current densities in AZ31 Mg alloys when compared to their cast, extruded and annealed counterparts, respectively. In contrast, Hosaka et al. [64] reported ECAP AZ31 alloy to be characterized by a corrosion rate of 87.5%, higher than its annealed counterpart, which they credit to the high amount of dislocations. In their study, ECAP process has been carried out below the recrystallization temperature of Mg (240°C). Hosaka et al. [64] also increased the ECAP temperature from 200°C to 300°C, which substantially improved the corrosion performances of ECAP AZ31. The detrimental effects of dislocations have been reported in other studies as well [69], where the effects of shot peening on the polarization curves of cold rolled and annealed AZ31 in 0.9% solution NaCl were investigated. A corrosion current density of 19 and 81 $\mu A/cm^2$ was observed for the nontreated and for the severely shot peened alloy, respectively. In addition, different process parameters have been reported to influence the corrosion behavior [70]. Bertolini et al. [71] reported reducing the feed rate from 0.1 to 0.01 mm/rev to reduce the corrosion potential from −1.79 to −1.69 V, the corrosion current density to 50% and the corrosion rate to about 30%.

AZ alloys are reported to possess a better corrosion behavior compared to pure Mg, but not to high-pure Mg. Mueller et al. found that AZ31 alloys are

TABLE 3.7

Summary of the corrosion rates for AZ series Mg alloys.*

Alloy	Additional elements	Corrosive environment	Immersion time	Corrosion rate (mm/year)	Procedure	Ref.
AZ61	—	SBF, 37°C	7 days	1.8	WL	[54]
AZ61	1% Y	SBF, 37°C	7 days	0.72	WL	[54]
AZ61	2% Y	SBF, 37°C	7 days	0.29	WL	[54]
AZ61	3% Y	SBF, 37°C	7 days	0.46	WL	[54]
AZ61	4% Y	SBF, 37°C	7 days	1.05	WL	[54]
AZ31 (extruded)	—	HBSS, 37°C	7 days	0.83	WL	[59]
AZ31 (extruded)	—	HBSS, 37°C	14 days	1.06	WL	[59]
AZ31 (extruded)	—	HBSS, 37°C	21 days	1.39	WL	[59]
AZ31 (extruded)	—	HBSS, 37°C	28 days	1.68	WL	[59]
AZ31	—	EBSS, 37°C	7 days	0.79	WL	[25]
AZ31	—	EBSS, 37°C	14 days	0.67	WL	[25]
AZ31	—	EBSS, 37°C	21 days	0.55	WL	[25]
AZ31	—	MEM, 37°C	7 days	1.29	WL	[25]
AZ31	—	MEM, 37°C	14 days	1.02	WL	[25]
AZ31	—	MEM, 37°C	21 days	1.19	WL	[25]
AZ31	—	MEMp, 37°C	7 days	1.94	WL	[25]
AZ31	—	MEMp, 37°C	14 days	1.29	WL	[25]
AZ31	—	MEMp, 37°C	21 days	0.94	WL	[25]
AZ31	—	Sub-cutis Lewis rats	7 days	0.34	WL	[25]
AZ31	—	Sub-cutis Lewis rats	14 days	0.34	WL	[25]
AZ31	—	Sub-cutis Lewis rats	21 days	0.22	WL	[25]
AZ31 (annealed)	—	RPMI, 37°C	24 h	1.56	WL	[64]
AZ31 (extruded)	—	RPMI, 37°C	24 h	1.76	WL	[64]
AZ31 (ECAP)	—	RPMI, 37°C	24 h	3.1	WL	[64]
AZ31 (Fe/ Mn=0.05)	—	SBF 5×, 37°C	72 h	16.4	WL	[61]
AZ31 (ECAP; Fe/Mn=0.05)	—	SBF 5×, 37°C	72 h	5.9	WL	[61]
AZ31 (wrought)	—	Hank's, 37°C	31 days	0.26	WL	[65]
AZ31 (hot rolled)	—	DMEM, 37°C	14 days	0.06	H	[62]
AZ31 (extruded)	—	Hank's, 37°C	30 days	0.15	WL	[63]

* All AZ samples are cast, unless otherwise specified. Additional information can be found on www.routledge.com/9780367429454.

characterized by lower corrosion current densities than pure Mg (99.98% pure) in different corrosive environments. Abidin et al. [21] demonstrated that Mg alloys including ZE41 and AZ91 corrode faster than high-purity Mg in Hank's solution at 37°C. The corrosion resistance of AZ alloys is better than that of pure Mg and their mechanical properties are appropriate for their application in orthopedic devices. Yet, their mechanical properties degrade rapidly and are considered inappropriate after three months, which is commonly understood as a sufficient healing time. Adekanmbi et al. [72] reported that after three months of immersion in phosphate-buffered saline at 37°C, the yield and tensile strength are 63 and 74 MPa, respectively, lower than those of cortical bones (Table 1.1, Chapter 1). This, in addition to the toxic effect of Al, has led researchers to study other alloys as reported subsequently.

3.3.1.3   AZ Alloys: Biocompatibility

The compatibility of a material can be assessed by means of coagulation and platelet aggregation for cardiovascular applications, and by either direct or indirect cell culture studies for orthopedic applications. In Table 3.8, the bio-compatibility studies available in literature for both these applications are summarized. The results are related to AZ91 and, particularly, AZ31 alloys, since very few studies can be found on other AZ series alloys [73,74]. AZ alloys, as well as all other Al-based alloys, are not directly applicable for bio-medical implants due to the long-term toxic effects of Al.

Mochizuki found that AZ31 alloys yield cell viability comparable to that of pure magnesium, with HUVEC-type cells being even higher. However, doubts on the cytocompatibility of AZ alloys remain. All studies related to L-929 cells show cell viability lower than 70% (Table 3.8). The limited cytocompat-ibility is due to corrosion cracks and hydrogen release that do not provide an adequate environment for cells to survive [77]. When corrosion products fall out, cell attachment on the material's surface becomes difficult [76]. In add-ition, the H_2 release can remove the attached cells from the surface and leads to an increase in the pH, thereby inducing cell death [75]. The environment around corrosion pits is particularly harmful for cells [78], since the pH can rise to levels above 8.5 which, in addition to the resulting high local concen-tration of Mg^{2+} ions, hampers cell attachment and proliferation (as shown, for example, on human bone marrow stromal cells (hBMSCs) in [79]). Cell cycle is widely reported to be influenced by the osmolarity: cell proliferation is promoted by osmotic swelling, whereas delayed in hyper-osmotic solutions [80]. Wong et al. [81] reported that Mg ion concentration of 50 ppm could stimulate osteogenic differentiation, whereas downregulation of osteogenesis-related genes was observed at a concentration of 200 ppm.

3.3.2   AM Alloys

AM alloys are characterized by Al and Mn as main alloying elements and they were initially developed to increase the manufacturability of AZ alloys.

TABLE 3.8

Summary of the biocompatibility of AZ31 and AZ91 alloys.*

		Cell viability					
Alloy	Test environment	Cell type, procedure	Time of culture	Result (%)	Platelet adhesion	Hemolysis (%)	Ref.
AZ31 (annealed)	DMEM, 37°C	L6, DC	1 day	88.6	—	—	[61]
AZ31 (annealed)	DMEM, 37°C	L6, DC	2 days	86.8	—	—	[61]
AZ31 (annealed)	DMEM, 37°C	L6, DC	3 days	92.9	—	—	[61]
AZ31 (ECAP)	DMEM, 37°C	L6, DC	1 day	89.4	—	—	[61]
AZ31 (ECAP)	DMEM, 37°C	L6, DC	2 days	96.7	—	—	[61]
AZ31 (ECAP)	DMEM, 37°C	L6, DC	3 days	91.4	—	—	[61]
AZ91	DMEM, 37°C	L-929, DC	2 days	50	—	—	[75]
AZ91	DMEM, 37°C	L-929, DC	5 days	55	—	—	[75]
AZ91	DMEM, 37°C	L-929, DC	7 days	58	—	—	[75]
AZ31 (extruded)	DMEM, 37°C	MG63, IC	1 day	72.1	—	—	[68]
AZ31 (extruded)	DMEM, 37°C	MG63, IC	2 days	83.9	—	—	[68]
AZ31 (extruded)	DMEM, 37°C	MG63, IC	3 days	77	—	—	[68]
AZ31 (ECAP)	DMEM, 37°C	MG63, IC	1 day	81.8	—	—	[68]
AZ31 (ECAP)	DMEM, 37°C	MG63, IC	2 days	74.5	—	—	[68]
AZ31 (ECAP)	DMEM, 37°C	MG63, IC	3 days	80.9	—	—	[68]
AZ31 (ECAP +BP)	DMEM, 37°C	MG63, IC	1 day	79.7	—	—	[68]
AZ31 (ECAP +BP)	DMEM, 37°C	MG63, IC	2 days	83.3	—	—	[68]
AZ31 (ECAP +BP)	DMEM, 37°C	MG63, IC	3 days	86.7	—	—	[68]
AZ91	DMEM, 37°C	L-929, DC	2 days	50.1	—	—	[76]
AZ91	DMEM, 37°C	L-929, DC	2 days	55.6	—	—	[76]
AZ91	DMEM, 37°C	L-929, DC	2 days	59.0	—	—	[76]
AZ31	DMEM, 37°C	Primary human osteo-blast, IC	1 day	106.5	—	—	[69]
AZ31	DMEM, 37°C	Primary human osteo-blast, IC	3 days	100.2	—	—	[69]
AZ31	DMEM, 37°C	Primary human osteo-blast, IC	7 days	121.8	—	—	[69]
AZ31 (SSP)	DMEM, 37°C	Primary human osteo-blast, IC	1 day	91.2	—	—	[69]
AZ31 (SSP)	DMEM, 37°C	Primary human osteo-blast, IC	3 days	80.2	—	—	[69]

(Continued)

TABLE 3.8 (Cont.)

Alloy	Test environment	Cell viability			Platelet adhesion	Hemolysis (%)	Ref.
		Cell type, procedure	Time of culture	Result (%)			
AZ31 (SSP)	DMEM, 37°C	Primary human osteo-blast, IC	7 days	112	—	—	[69]
AZ31	α-MEM, 37°C	MC3T3-E1, DC	1 day	74.3	—	—	[77]
AZ31	α-MEM, 37°C	MC3T3-E1, DC	3 days	55.2	—	—	[77]
AZ31 (annealed)	DMEM, 37°C	L6, DC	1 day	88.6	—	—	[61]
AZ31 (annealed)	DMEM, 37°C	L6, DC	2 days	86.8	—	—	[61]
AZ31 (annealed)	DMEM, 37°C	L6, DC	3 days	92.9	—	—	[61]
AZ31 (ECAP)	DMEM, 37°C	L6, DC	1 day	89.4	—	—	[61]
AZ31 (ECAP)	DMEM, 37°C	L6, DC	2 days	96.7	—	—	[61]
AZ31 (ECAP)	DMEM, 37°C	L6, DC	3 days	91.4	—	—	[61]

* All the samples are cast, unless otherwise specified. Additional information can be found on www.routledge.com/9780367429454.

Luo et al. [82] reported that AZ alloys are characterized by a low extrudability due to the formation of ternary Mg–Al–Zn phase particles at a low temperature (338°C) and showed that the extrudability of AM30 alloy is higher than that of AZ31. The increasing interest in AM alloys is therefore comprehensible, where particularly AM50 and AM60 shall be mentioned here. However, their yield and tensile strengths are lower than AZ alloys. In addition, an increase in the amount of Mn lowers the susceptibility to Fe impurities since a weight ratio Fe/Mn of 0.032 is considered the threshold for this ternary alloy [17–19]. The most likely mechanism explaining the moderation of Fe is its incorporation into an intermetallic AlMnFe compound, which is less active as a local cathode than $FeAl_3$ [39,83]. However, Mn, in concentrations higher than 10 μmol/l in the blood, has been shown to induce "Manganism," a neurological disorder similar to Parkinson's disease [84].

3.3.2.1 AM Alloys: Mechanical Properties

Mechanical properties such as tensile and compressive yield strength, tensile and compressive strength and elongation to failure of different AM alloys are gathered in Table 3.9. Because of the increase in the formability of AM alloys compared to the AZ alloys, several studies have been carried out by researchers in aerospace and automotive field, especially aiming to render their mechanical properties comparable to those of AZ alloys. Several studies have thus dealt with the addition of other alloying elements, especially RE elements [85–89]. In fact, it is commonly accepted that the RE elements are effective to improve the mechanical properties of Mg–Al-based alloys because of the effect of purifying alloy and grain refinement [90,91]. Braszczyńska-Malik added RE elements on die-casted AM50 alloy, reporting that the addition of a RE element of 5 wt.% determines an increment of 9% and 12% in the yield strength and UTS, respectively [92]. Su et al. [93] investigated the effect of yttrium on AM60 alloy, and stated its tensile strength to have increased from 179 to 192 MPa by adding a 0.9 wt.% of yttrium, due to the formation of Al_2Y phases or the grain refinement effect of yttrium. Moreover, in the same work, they assessed the effect of different manufacturing processes, reporting hot rolled components to be characterized by higher mechanical properties than their cast counterparts due to the grain refinement induced by the rolling process. In addition, Ding et al. [94] demonstrated the mechanical properties of AM alloys to be affected not only by the process but also by the parameters of the process itself.

From Table 3.9 it can be noticed that by increasing the amount of Al, either the yield or the tensile strength increases. This agrees with the study of Yu et al. [96], where they compared the mechanical behavior of several cast AM alloys, such as AM01, AM11, AM31, AM61 and AM91 (Figure 3.4).

They ascribed these results to the strengthening induced by Al due to its grain refinement effect and either to its solid solution strengthening or precipitation hardening. However, the occurrence of high level of precipitations induced a decrease in the ductility (Al content higher than 3 wt.%). This behavior is comparable to that observed in AZ alloys, where increasing the Al content first increase the mechanical properties, but a further increment does not cause a remarkable change and, when 9 wt.% Al is added, the UTS and yield strength decrease. Again, elongation to failure decreases by adding more than 2 wt.% Al [49].

3.3.2.2   AM Alloys: Corrosion Resistance

Unlike AZ alloys, little research has been done about the corrosion behavior of AM alloys in a physiologically relevant environment. Anawati et al. [97] studied the polarization curves of AM60 alloy and the effect of adding 1 wt.% and 2 wt.% of Ca as alloying element in a 0.9% NaCl solution at 37°C. AM60 alloy exhibited a corrosion potential of −1.48 V; this potential was substantially increased to a nobler potential of −1.44 V by the addition of 1 wt.% Ca, but shifted back to −1.47 V when the Ca content was increased to 2 wt.%. This shift of corrosion potential resulted in an increase of the corrosion current density. The corrosion current density of the AM60 alloy increased slightly from 1.16×10^{-5} to 1.44×10^{-5} A/cm^2 with the addition of 1 wt.% Ca in the alloy, and further to 2.67×10^{-5} A/cm^2 in the 2 wt.% Ca content case. Anawati et al. attributed the increase in the current density by adding Ca to the higher volume fraction of the intermetallic phases. Abdal-hay et al. [98] reported die-cast AM50 to be characterized by a corrosion potential of −1.11 V and a corrosion current density of 9.74 μA/cm² in SBF at 37.5°C. However, no studies deal with the comparison between the corrosion behavior of AM and AZ alloys in a physiologically relevant environment. However, several studies have

TABLE 3.9
Summary of AM alloys mechanical properties.*

Alloy	Additional element	Yield strength (MPa)	UTS (MPa)	Elongation to failure (%)	Compressive yield strength (MPa)	Ultimate compressive strength (MPa)	Ref.
AM60 (die-casted)	—	—	156.1	6.1	—	—	[85]
AM60 (die-casted)	0.5% Ce	—	168	8.3	—	—	[85]
AM60 (die-casted)	1% Ce	—	196.2	9.7	—	—	[85]
AM60 (die-casted)	1.5% Ce	—	188.1	9.3	—	—	[85]
AM20	—	160	202	7	—	—	[95]
AM50 (hot rolled)	—	241	305	14.8	—	—	[94]
AM50 (die-casted)	1% RE	132.2	224	—	—	—	[92]
AM50 (die-casted)	3% RE	139	235.1	—	—	—	[92]
AM50 (die-casted)	5% RE	144.8	248.1	—	—	—	[92]
AM11	—	42.1	142.8	12.3	—	—	[96]
AM31	—	60.1	184.6	14.8	—	—	[96]
AM61	—	87.5	201.4	10.7	—	—	[96]
AM91	—	122.9	180.6	5.6	—	—	[96]
AM30 (extruded)	—	170	240	12	—	—	[82]
AM60	—	56	179	11.8	—	—	[93]
AM60	0.3% Y	57	185	12.5	—	—	[93]
AM60	0.6% Y	59	186	12.4	—	—	[93]
AM60	0.9% Y	62	192	12.6	—	—	[93]
AM60 (hot rolled)	—	221	293	10.3	—	—	[93]
AM60 (hot rolled)	0.3% Y	249	300	13.3	—	—	[93]
AM60 (hot rolled)	0.6% Y	231	314	15.4	—	—	[93]
AM60 (hot rolled)	0.9% Y	255	303	17.1	—	—	[93]
AM50	—	67	140	—	87	304	[86]
AM50	3% RE	78.1	151.1	—	91	319	[86]
AM50	5% RE	80.1	153	—	94	326.8	[86]
AM31	—	173	236.3	18.4	—	—	[87]
AM31 (TRC)	—	200.9	256.7	20.3	—	—	[87]
AM31	Ca	161.9	221.4	20.7	—	—	[87]
AM31 (TRC)	Ca	238.1	302.3	25	—	—	[87]
AM31	MM	170.2	233.5	25.6	—	—	[87]
AM31 (TRC)	MM	252.1	309.8	24.2	—	—	[87]
AM31	Y	202.8	258.6	21.1	—	—	[87]
AM31 (TRC)	Y	262.3	316.3	22.1	—	—	[87]
AM31 (DSR)	—	195.1	264.9	17.6	—	—	[87]
AM31 (TRC + DSR)	—	221.6	282.6	18.4	—	—	[87]

(Continued)

TABLE 3.9 (Cont.)

Alloy	Additional element	Yield strength (MPa)	UTS (MPa)	Elongation to failure (%)	Compressive yield strength (MPa)	Ultimate compressive strength (MPa)	Ref.
AM50	—	87.2	145.2	2.8	103.8	294.1	[88]
AM50	0.5% Ce	93.2	156.7	3.7	108.9	303	[88]
AM50	1% Ce	97.7	167.6	4.8	115.9	314.1	[88]
AM50	—	74.9	152.4	3.3	—	—	[89]
AM50	0.3% Y	85.5	177.4	3.4	—	—	[89]
AM50	0.3% Ce	88.4	187	3.6	—	—	[89]
AM50	0.3% Y + 0.3% Ce	93.2	215.2	4.1	—	—	[89]
AM50	0.3% Y + 0.6% Ce	117.5	244.9	4.3	—	—	[89]
AM50	0.3% Y + 0.9% Ce	97	196.6	3.4	—	—	[89]
AM50	0.3% Y + 1.2% Ce	80.7	174.2	3.2	—	—	[89]

* All the AM samples are cast, unless otherwise specified. Additional information can be found on www.routledge.com/9780367429454.

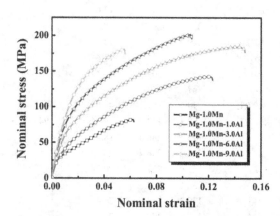

FIGURE 3.4 Effect of aluminum on tensile tests of different AM alloys. Reprinted with permission from Taylor & Francis [96].

been conducted at room temperature in NaCl solutions. Although these studies have been carried out for automotive and other industrial applications, the results can be considered to hold also in the biomedical field. Bender et al. [99] studied the polarization curves of die-cast AM20 and AZ91 alloy and of wrought AZ31 alloy in NaCl solution (0.01 M) at 23°C. They reported the AZ91 alloy to be characterized by the lowest corrosion current density (8 μA/cm^2), followed by AM20 alloy (10 μA/cm^2) and AZ31 (14 μA/cm^2) alloy. However, due to the difference in the manufacturing process, the AZ31 alloy cannot be compared. Zhao et al. [100] studied the corrosion rate for cast AZ91 and AM60 alloy and for extruded AZ31 and AM30. In both the cases, AZ alloys were characterized by a lower corrosion rate (Table 3.10).

The higher corrosion rates for the AM alloys compared with the AZ alloys could be due to (i) the AM alloys having a beta-phase that is more effective as a cathode for the hydrogen evolution reaction or (ii) the AZ having a surface film more resistant to microgalvanic corrosion compared with the AM alloys. In addition, the effect of RE addition as an alloying element has also been studied [101–103]. Mert et al. studied the impact of Ce addition to AM50 and found out that the corrosion resistance is improved with a larger Ce content due to the formation of the $Al_{11}Ce_3$ phase and reduction of the $Mg_{17}Al_{12}$ phase [104].

3.3.2.3 AM Alloys: Biocompatibility

The biocompatibility of AM alloys has not been widely investigated. At the best of the authors' knowledge, only one research paper has dealt with this topic. Abdal-hay et al. [98] studied the biocompatibility of die-cast AM50 alloy. In particular, they assessed the cell proliferation of MC3T3-E1 after 1, 3 and 5 days of culturing using MTT assay, and the results are reported in Figure 3.5.

3.3.3 Mg–Al–RE Alloys

Because of the high-melting point of RE elements, Mg–Al–RE alloys have been developed to improve the poor mechanical properties of AZ and AM alloys at temperature above 125°C in automotive applications [105,106]. However, due to the presence of aluminum, their assessment in a physiologically relevant environment has been rarely carried out. The majority of the works available in literature deal with their high-temperature properties. These alloys,

TABLE 3.10
Corrosion rate for different AM and AZ alloys obtained with two different manufacturing processes

Alloy	1-day immersion (mL/cm^2/day)	7 days immersion (mL/cm^2/day)	12 days immersion (mL/cm^2/day)
AZ31 (extruded)	1.2	1.0	1.0
AM30 (extruded)	1.9	3.8	3.9
AZ91 (cast)	1	2	3
AM60 (cast)	1.8	4.1	4.8

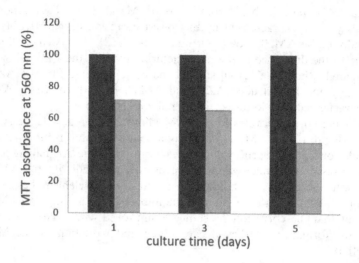

FIGURE 3.5 Cell proliferation of MC3T3-E1 after different culturing days using MTT assay. Modified from [98].

in fact, exhibit major improvement in creep resistance due to the complete suppression of $Mg_{17}Al_{12}$ phase and the formation of the highly thermal stable Al–RE intermetallic compounds, $Al_{11}RE_3$ and Al_2RE. Typical Mg–Al–RE alloys are divided into AE and LAE alloys. The former are characterized by aluminum and RE as main alloying elements, whereas the latter by lithium, aluminum and RE as main alloying elements. RE elements added in Mg–Al–RE alloys are usually Ce-rich misch metals (MMs) whose typical composition is 52–55 wt.% Ce, 23–25 wt.% La, 16–20 wt.% Nd and 5–6 wt.% Pr. However, because of the large consumption of Ce-rich MMs, used in Nd (Pr)–Fe–B permanent magnets, and the subsequently rising price of Nd and Pr at present, the price increase of Ce-rich MM is inevitable. Thus, Zhang et al. [107] tried to substitute Ce-rich MM with the mixture of Ce and La in AE alloys, reporting their modified AE44 to have higher mechanical properties than its commercial counterpart: the yield strength increased from 147 to 160 MPa, the UTS from 247 to 270 MPa and the elongation from 11% to 13%. In addition, mechanical properties have been improved by substituting Ce-rich MM with the mixture of Ce and La also at high temperatures due to the higher thermal stability of $Al_{11}(Ce,La)_3$ than that of $Al_{11}RE_3$ [108].

3.3.3.1 Mg–Al–RE Alloys: Mechanical Properties

Because of their increased mechanical properties at high temperature, most of the works dealing with Mg–Al–RE alloys are about their creep resistance and their mechanical properties at high temperature. In Table 3.11, a summary of the mechanical properties at room temperature of these alloys is gathered. In the column RE elements, "MM" stands for the addition of Ce-rich MM.

TABLE 3.11
Summary of Mg–Al–RE alloys mechanical properties.*

Alloy	RE elements	Yield strength (MPa)	UTS	Elongation to failure (%)	Compressive yield strength (MPa)	Ultimate compressive strength (MPa)	Ref.
AE44 (HPDC)	MM	147	247	11	—	—	[107,109]
AE41 (HPDC)	Ce–La mixture	128	233	11	—	—	[107]
AE42 (HPDC)	Ce–La mixture	137	240	11	—	—	[107]
AE44 (HPDC)	Ce–La mixture	160	270	13	—	—	[107]
AE46 (HPDC)	Ce–La mixture	173	261	8	—	—	[107]
AlLa41	Lanthanum	133	236	12	—	—	[109]
AlLa42	Lanthanum	140	245	13	—	—	[109]
AlLa44	Lanthanum	155	265	12	—	—	[109]
AlLa46	Lanthanum	171	257	7	—	—	[109]
AE42 (permanent mold cast)	MM	97	202	12	—	—	[110]
AE42 (HPDC)	MM	133	228	8.9	—	—	[111]
AE42 (die-cast)	MM	120.1	239.6	11.8	—	—	[112]
AE44 (die-cast)	MM	131.1	252.1	13.3	—	—	[112]
AE44 (die-cast)	Ce–La mixture	127.7	258.3	14.3	—	—	[112]
AE42 (HPDC)	MM	101	184	5	—	—	[113]
AE42 (HPDC)	MM	141.9	232.2	12	—	—	[114]

* All the Mg-Al-RE samples are cast, unless otherwise specified.

Although its price is increasing as mentioned above, its use remains the most frequent way to introduce RE as alloying elements. However, some studies have started substituting Ce-rich MM not only with Cerium and Lanthanum but also with Neodymium. In addition, several studies have been carried out on high-pressure die-cast (HPDC) Mg–Al–RE alloys since this manufacturing process is extensively used for automotive applications.

Both the yield and tensile strength of Mg–Al–RE alloys are lower than AM and AZ alloys, whereas the elongation to failure is higher than AZ alloys and slightly better than AM alloys [107,109–114]. All the Mg–Al–RE samples are cast, unless otherwise specified. Additional information can be found on www.routledge.com/9780367429454.

3.3.3.2 Mg–Al–RE Alloys: Corrosion Resistance

The diffusion of Mg–Al–RE alloys in the biomedical studies is still limited. The majority of the studies deals with LAE alloys, whereas only one work has been found concerning AE alloys. In this work, Minárik et al. [115] investigated the corrosion behavior of extruded AE21 and AE42 alloys in 0.1 M NaCl solution, assessing also any positive effects of ECAP process. While extruded AE21 and AE42 are characterized by the same polarization resistance, the ECAP process resulted in an opposite behavior of the two alloys: after ECAP process, the polarization resistance of AE21 has been reported to be inferior to that of AE42 due to a different chemical composition. While Al and RE particles are concentrated in stripes in the extruded alloys, ECAP process leads to their fragmentation and these fragmented particles have been reported to be better distributed in AE42 alloy, improving their corrosion resistance because better spatial distribution leads to a better stabilization of the $Mg(OH)_2$ layer [115]. The same improvement in corrosion resistance due to a better distribution of Al-rich secondary phases after ECAP process has also been reported for LAE442 in different physiologically relevant environments (Figure 3.6) [116]. The results agree with those found by Minárik et al. [117], who reported a higher corrosion resistance of LAE442 after ECAP compared to the extruded counterpart.

Leeflang et al. [118] then investigated the long-term biodegradation properties of Mg–Li–Al–RE alloys in HBBS solution at 37°C. They assessed the hydrogen evolution of LA92, LAE912 and LAE922, comparing the results with WE43, an Mg–Y–RE alloy (Figure 3.7), that has been subjected to animal and preclinical experiments [119].

Leeflang et al. attributed the better corrosion behavior of LA92 to the presence of lithium that shifted the pH value of the solution to a higher value, thereby stabilizing the $Mg(OH)_2$ layer on the surface [15]. The addition of RE elements to the Mg–Li–Al alloys to form the LAE912 and LAE922 alloys, leads to the formation of intermetallic compounds of considerably large volume fractions, such as $Mg_{12}La$, $Mg_{12}Ce$ and Mg_3Nd, that have significantly different electrochemical properties, leading to localized corrosion and thus accelerating the entire degradation process. These results disagree with those found by Luo et al. [55], which reported a concentration of 0.3 wt.% of yttrium to decrease the corrosion rate of AZ91. The influence of RE elements is thus dependent on the composition of the base alloy. In addition, the corrosive environment has been reported to highly influence the corrosive responses of Mg alloys. Comparing LAE442 and AZ91 alloys, Witte et al. reported AZ91 to be characterized by a lower *in vitro* corrosion rate (2.8 instead of 6.9 mm/year), but higher if *in vivo* conditions are considered, that is 3.5 ×

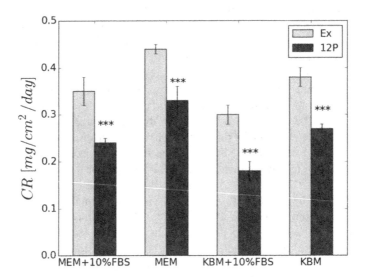

FIGURE 3.6 Corrosion rate of LAE442 calculated from mass loss (mg/cm^2/day) after 14 days in 40 ml of media (KBM or MEM, with or without 10% FBS). Ex = extruded; 12P = 12 ECAP passes. Reprinted with permission from Elsevier [116].

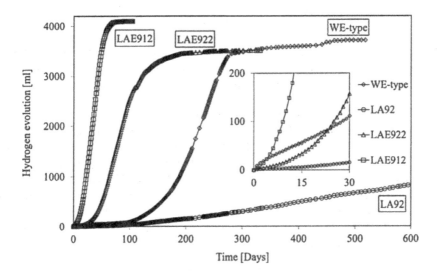

FIGURE 3.7 Degradation profiles of the LA92, LAE912 and LAE922 alloys in Hank's solution in comparison with the profile of the WE-type alloy. The inset shows the different initial responses of the alloys to Hank's solution over a period of the first 30 days. Reprinted with permission from Elsevier [118].

10^{-4} mm/year against 1.2×10^{-4} mm/year [120], agreeing with that reported in a previous paper [14], where the *in vivo* corrosion behavior of AZ31, AZ91, WE43 and LAE442 alloys have been assessed after 6 and 18 weeks of implantation in guinea pigs. The results reported LAE442 to be characterized by the lower corrosion rate, since the addition of RE elements is known to minimize the aluminum threshold for effective corrosion protection [121]. Furthermore, addition of lithium is known to alkalize the corrosion layer and therefore stabilize the formed magnesium hydroxides within the corrosion layer [14]. Different corrosion behaviors in different corrosive environments have also been found by Mueller et al. [122]. Studying the polarization curves for AZ31 and LAE442 in different environments, they could not define which among them is characterized by the higher corrosion resistance. Since the remaining works found by the authors still deal with the *in vivo* corrosion behavior of LAE442 [123–130], the authors felt to be appropriate to leverage on the studies carried out for automotive applications to provide further comparisons between the corrosion rates of AZ alloys (characterized by a better corrosion behavior than AM alloys) and Mg–Al–RE alloys. Berkmortel et al. [114], carrying out salt spray corrosion tests, reported AE42 to be characterized by a corrosion rate almost three times higher than AZ91, but lower than AM50. Moreover, Khabale and Wani [131] found AZ91 to be characterized by better tribological properties than AE42. In addition, LAE442 is reported to be characterized by a lower corrosion rate than AZ31 in 0.5% NaCl solution, but higher in 3.5% NaCl solution [132]. However, the different Al content does not allow a direct comparison of the results. Rosalbino et al. [133] compared AZ91 alloy with AE91 (with Ce as RE), reporting a better corrosion resistance of AE91 because of the higher stability of passive film. This result agrees with the higher polarization resistance found for Mg–6Al–4RE than AM60 by Liu et al. [134] and with Mert et al. [104] that reported adding Ce to improve the corrosion resistance of AM50 due to the formation of the $Al_{11}Ce_3$ phase and reduction of the $Mg_{17}Al_{12}$ phase.

3.3.3.3 Mg–Al–RE Alloys: Biocompatibility

Few works deal with the biocompatibility of Mg–Al–RE alloys, and they are all about LAE442 alloy, where the RE elements can be represented by Ce-rich MM, Cerium or Neodymium (Nd). Minárik et al. [116] studied the cell viability of L929 cells of extruded and ECAP LAE442 via indirect test in MEM with different amounts of FBS. The cell viability has been reported to be far above 70% limit, stated in ISO standard (dashed line in Figure 3.8), and no difference in cytotoxicity between the samples before and after ECAP processing has been reported.

The results agree with those obtained by Krämer et al. [129] and Zhou et al. [135]. The latter, besides assessing the VSMC and ECV304 cell viabilities in different Mg–Li–Al–RE-based alloys extraction medium solutions, evaluated the hemolysis ratio of different Mg–Li–Al–RE alloys, reporting that their hemolysis ratios were □5% (a judging criterion for excellent blood compatibility), except that of Mg–8.5Li–2Al–2RE, which was characterized by a hemolysis ratio of 6.5%. Other works available in literature then deal with *in vivo* biocompatibility

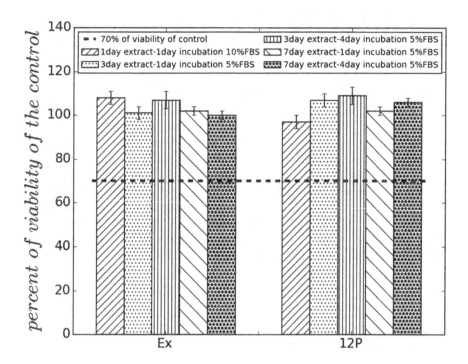

FIGURE 3.8 Metabolic activity of L929 after 1-day or 4-day exposition to LAE442 extracts prepared using various media and extraction periods. Dashed line stands for the limitary value of cytocompatibility, set as 70% of metabolic activity of the untreated control by ISO 10993-5. Reprinted with permission from Elsevier [116].

of LAE442 alloy [136]. Angrisani et al. [126] studied the biocompatibility of LAE442 implants for up to 3.5 years in rabbits, reporting a very slow degradation and a very good cellular compatibility of this alloy: close bone to implant contact and the absence of lytic areas around the implant and cavities within the cortical bone have been observed, representing most desirable features of degradable orthopedic implants. Furthermore, Rössig et al. [128] found a higher periosteal formation of new bone in LAE442 implants than in the standard austenitic steel (1.4441LA alloy) intramedullary nailing system. Concerning the substitution of Ce-rich MM with Ce and Nd, the results reported LAE442 to be the best option: while LACe442 is characterized by a far lower bone adhesion and new bone formation [137], LANd442 is reported to lack in mechanical strength [138].

3.4 MG–ZN ALLOYS

Zn has been considered as alloying element in biocompatible Mg alloys, since it is an essential trace mineral for hundreds of biological enzymes, being required by the human body at 15 mg/day [139,140] and it has a solubility limit of 6.2 wt.% [141]. Causing solid solution strengthening, Zn effectively improves the mechanical properties of magnesium: the strength of Mg has been reported to increase up to 280 MPa by adding 6 wt.% of zinc [142]. In addition, Zn has been reported either to enhance the tolerance limit or reduce the effects of the three main impurities (Fe, Cu and Ni) when their solid solubility limits are exceeded [143]. Song and Atrens [17] reported the Ni tolerance limit to be increased up to 20 ppm in Mg–Al–Mn alloys with an addition of 3 wt.% Zn, and it can further reduce the corrosion rate of ternary alloys when Ni and Fe tolerance levels are reached. However, at Zn concentration above 6 wt.%, second phases form and the corrosion resistance lowers [144], leading to localized corrosion [145]. Another advantage of Zn alloying is the decrease in hydrogen evolution along with the decrease in solubility of the Mg matrix. Both, Mg^{2+} ions and Zn^{2+} ions bind with free OH^- anions forming $Mg(OH)_2$ and $Zn(OH)_2$, respectively, and reducing the amounts of free H_2 [146]. However, cytotoxicity has been identified *in vitro* with exposure of cells to high concentrations of Zn [147,148]. The main drawback regarding Zn's biocompatibility is in fact the reaction of Zn^{2+} with hydrochloric acid (HCl). Zn^{2+} evolves from the oxidation reaction of Zn, used as alloying material,

$$Zn \rightarrow Zn^{2+} + 2e^- \qquad (3.1)$$

and HCl is reduced according to:

$$2HCl + 2e^- \rightarrow H_2 + 2Cl^- \qquad (3.2)$$

leading then to the formation of $ZnCl_2$, which is known to damage stomach parietal cells [149].

In recent years, different Zn-containing Mg alloys have been produced and investigated: Mg–Zn binary alloys and Mg–Zn ternary alloys, for example, Mg–Zn–Zr alloys, Mg–Zn–Ca alloys [included bulk metallic glasses (BMGs)], Mg–Zn–Mn alloys and Mg–Zn–RE alloys.

3.4.1 Mg–Zn Binary Alloys

As previously mentioned, to improve the corrosion resistance and mechanical properties of pure magnesium, different procedures have been studied, and adding alloying element is one of the most investigated. In the initial development period of Mg biomaterials, some commercial engineering Mg alloys have been investigated as biomaterials (e.g., AZ31, AZ91, AM60 and WE43). Unfortunately, it has been found that the administration of Al and RE may induce latent toxic and harmful effects on the human body [46,150,151]. Consequently, new alloying elements are being considered, and since Zn is an indispensable trace element in the human body, binary and ternary Mg–Zn-based alloys have started to draw researchers' attention.

3.4.1.1 Mg–Zn Binary Alloys: Mechanical Properties

Besides being a trace element, Zn has been deemed to be a good alloying element for biomedical applications, since it considerably improves the low strength of magnesium. Zn can, in fact, provide the advantage of solution strengthening and aging strengthening. Peng et al., studying the mechanical properties of Mg-xZn (x = 0.5, 1, 1.5, 2 wt.%) alloys, reported either the yield strength or the UTS to be maximum when the Zn content was the highest [152]. They attributed the improved strength to the solid solution strengthening. In addition, they stated the higher the Zn content, the finer the grain size, thus attributing the increase in the mechanical properties also to the Zn grain refinement effect. Cai et al. [153] studied as-cast Mg–Zn alloys with Zn content of 1, 5 and 7 wt.%. They found that increasing the Zn content leads to a reduction in the ductility (13.77% and 6% for Mg–1Zn and Mg–7zn, respectively). The tensile and yield strength increase until 5% of Zn due to grain refinement, solid solution strengthening and precipitation of MgZn second phases. The presence of second phases is also responsible for the reduced plasticity. On the one hand, second phases may hinder the dislocation reduction and increase the dislocation density, and on the other hand, the second phases dispersed at the grain boundary could be new crack source, which expands easily and eventually results in brittle failure. According to Cai et al., these second phases are responsible for the reduction in strength and for the further decrease in the elongation to failure obtained for Mg–7Zn. In fact, micrographic analyses of Mg–7Zn have reported second phases to form a network structure with dendritic segregation along grain boundaries, resulting in residual defects that reduce the strength and elongation of alloy. However, due to the high costs associated with long solutionizing and aging treatment, Mg–Zn ternary alloys have been studied. Besides Mg–Zn–X (X = Zr, RE, Ca and Mn), which will be treated in the following, the effect of adding Strontium (Sr) has also been widely investigated [154,155]. Sr behaves similarly to calcium and can be commonly found as a Ca substitute in biomedical compounds such as hydroxyapatite [156,157]. In addition, Sr is an effective grain refiner. However, divergent results are available in literature on the optimum Sr concentration. While Li et al. [155] reported that in Mg–1Zn–xSr (x = 0.2, 0.5, 0.8, 1 wt.%) the higher the Sr content, the higher the strength and the elongation, Cheng et al. [154] reported the best mechanical properties to be obtained with an Sr content of 0.2 wt.%. In Table 3.12, the mechanical properties available in literature for the binary Mg–Zn system as well as for minor Mg–Zn–X ternary system are listed.

3.4.1.2 Mg–Zn Binary Alloys: Corrosion Resistance

As previously mentioned, Zn has the ability to reduce the corrosion enhancing effects of all of the common impurities including iron, nickel and copper. In addition, Zn has been reported to increase the corrosion resistance due to its high solid solubility limit. Guan et al. [158] stated that the solid solution of Zn in the α-Mg matrix phase increases the electric potential of the matrix and

TABLE 3.12

Summary of Mg–Zn alloys mechanical properties.*

Zn content (wt. %)	Additional element	Yield strength (MPa)	UTS (MPa)	Elongation to failure (%)	Compressive yield strength (MPa)	Ultimate compressive strength (MPa)	Ref.
1	—	25	130	18	—	—	[24]
1	—	20	102	7	—	—	[145]
1	—	61	188	14	—	—	[153]
4	—	58	217	16	—	—	[145]
5	—	76	195	8.5	—	—	[153]
7	—	67	136	6	—	—	[153]
1 (rolled)	—	160	239	7.2	—	—	[24]
0.5	—	38	95	4.2	32	137	[152]
1	—	42	99	6.1	35	140	[152]
1.5	—	51	109	5.9	42	145	[152]
2	—	65	121	5.3	54	155	[152]
0.5 (backward extruded)	—	62	145	17.2	49	237	[152]
1 (backward extruded)	—	91	169	18.7	64	295	[152]
1.5 (backward extruded)	—	101	190	17.2	65	305	[152]
2 (backward extruded)	—	111	198	15.7	74	315	[152]
5	—	120	212	10	—	—	[154]
5	0.2 wt.% Sr	117	233	15	—	—	[154]
5	0.6 wt.% Sr	115	215	13	—	—	[154]
5	1 wt.% Sr	107	194	9	—	—	[154]
1 (backward extruded)	—	89	187	11	82	241	[155]
1 (backward extruded)	—	93	211	11.8	90	247	[155]
1 (backward extruded)	—	117	210	11.5	111	257	[155]
1 (backward extruded)	—	130	249	12.6	129	278	[155]
6 (extruded)	—	169.5	279.5	18.8	—	433.7	[145]

* All the Mg–Zn samples are cast, unless otherwise specified. Additional information can be found on www.routledge.com/9780367429454.

improves the corrosion resistance. However, there are disagreements over the optimal Zn addition. Cai et al. reported that, despite the presence of the potentially detrimental MgZn intermetallic, increasing the Zn content from 1 to 5 wt.% leads to a decrease in the corrosion rate from 0.53 to 0.26 mm/year due to the formation of a passivation film on the surface [153]. However, excessive addition of Zn over 7 wt.% results in a network structure of MgZn intermetallic as a cathode, causing microgalvanic corrosion acceleration. These results are in contrast to those obtained by Zhang et al. [159]. They found that the corrosion rate increases by increasing the Zn content, slightly at Zn content lower than 4 wt.% and rapidly above this value. The differences present in literature can be related to the manufacturing process. While as-cast Mg–5Zn has been reported to be characterized by second phases, Zhang et al. [145] reported extruded Mg–6Zn alloy to consist of a uniform single phase after solid solution treatment and hot rolling. Different heat treatments have then been showed to have different results on the corrosion behavior of Mg–3Zn. Liu et al. [160], in fact, reported the solution treatment to enhance the corrosion resistance, whereas aging treatment decreases it. In Tables 3.13 and 3.14, the corrosion behavior of Mg–Zn binary alloys has been gathered in terms of polarization curves (Table 3.13) and corrosion rates (Table 3.14), highlighting, when performed, the manufacturing processes and the heat treatments which the alloys were subjected to.

The corrosion rate of Mg–Zn alloys has been compared to the corrosion rate of other biocompatible Mg alloys. Zang et al. [161] reported pure magnesium to be characterized by a corrosion rate more than double than that of extruded Mg–6Zn after three days of immersion in SBF at 37°C. These results agree with those obtained by Song [15]. He tested different alloys in Hank's solution at 37°C, reporting Mg–1Zn to be characterized by a corrosion rate of 6.22 mm/year, lower than that of 577 mm/year of commercially pure magnesium. However, AZ91 and high-pure Mg have been reported to be characterized by a lower corrosion rate, that is, 1.51 mm/year and 0.18 mm/year, respectively.

3.4.1.3   Mg–Zn Binary Alloys: Biocompatibility

Being an important macronutrient for humans and being involved in a wide range of physiological functions including protein synthesis, immune system regulation and many enzymatic reactions, Zinc has been considered a good biocompatible material with a very low cytotoxicity. This consideration is supported by the results available in literature, either by *in vivo* or *in vitro* studies. Concerning the latter, the *in vitro* hemolysis rate of Mg–6Zn has been found to be 3.4% [161], which is lower than the safe value of 5% according to ISO 10993-4, indicating that Mg–6Zn alloy exhibits good biocompatibility *in vitro*. These results agree with those obtained by Cui et al. [162], where the hemolysis rates of Mg–4.0Zn–1.5Sr alloy after 1, 3 and 5 days of immersion were always lower than the threshold of 5%, being 1.04%, 0.8% and 4.56%, respectively. In addition, they assessed the cytotoxicity of the material via the relative growth rate of cells according to the United States Pharmacopeia [163],

TABLE 3.13

Summary of the DC polarization curve results for Mg–Zn binary alloys.*

Zn content (wt. %)	Additional elements	Corrosive environment	E_{corr} (V)	i_{corr} (µA/cm²)	Ref.
0.5	—	SBF, 37°C	−1.6	53.3	[152]
1	—	SBF, 37°C	−1.58	178	[152]
1.5	—	SBF, 37°C	−1.57	374	[152]
2	—	SBF, 37°C	−1.58	423	[152]
0.5 (backward extruded)	—	SBF, 37°C	−1.57	20.1	[152]
1 (backward extruded)	—	SBF, 37°C	−1.56	47.4	[152]
1.5 (backward extruded)	—	SBF, 37°C	−1.53	57.1	[152]
2 (backward extruded)	—	SBF, 37°C	−1.50	59.3	[152]
5	—	0.9% NaCl, 37°C	−1.57	44	[154]
5	0.2 wt. % Sr	0.9% NaCl, 37°C	−1.56	34.9	[154]
5	0.6 wt. % Sr	0.9% NaCl, 37°C	−1.57	43.5	[154]
5	1 wt. % Sr	0.9% NaCl, 37°C	−1.56	41.3	[154]
5	—	Hank's, 37°C	−1.56	33.8	[154]
5	0.2 wt. % Sr	Hank's, 37°C	−1.51	22.4	[154]
5	0.6 wt. % Sr	Hank's, 37°C	−1.53	31.6	[154]
5	1 wt. % Sr	Hank's, 37°C	−1.54	26.6	[154]
6 (extruded)	—	SBF, 37°C	−1.82	237.5	[23]
1	—	SBF, 37°C	−1.82	67.3	[24]
1 (rolled)	—	SBF, 37°C	−1.81	40.8	[24]
1	—	Hank's, 37°C	−1.61	10.5	[24]
1 (rolled)	—	Hank's, 37°C	−1.55	7.5	[24]
6 (extruded)	—	SBF, 37°C	−1.62	45	[161]

* All the Mg–Zn samples are cast, unless otherwise specified.

reporting a zero level of cytotoxicity, in accordance with the results obtained in ref. [164] for Mg–2Zn and in refs. [145,165] for Mg–6Zn alloys, respectively. However, Gu et al. [24] reported cytotoxic effects of cast and rolled Mg–1Zn alloys, reporting hemolysis rates of 34.2% and 14.3%, respectively. However, dealing with the indirect cell viability of different cells, such as L-929, NIH3T3, MC3T3-E1, ECV304 and VSMC, they did not find any cytotoxic effects, agreeing with the results obtained by Seyedraoufi and Mirdamadi [166] for MG63 cells in Mg–2Zn extracts, Li et al. for hBMSCs cultured on Mg–6Zn [167] and Li et al. [168] and [155] for L-929 in extracts of Mg–Zn alloys containing different amounts of Sr. In these works, Sr has been reported to increase the biocompatibility of Mg–Zn alloys, and Nguyen et al. [169] reported that adding 1 wt.% of Sr on Mg–4Zn alloys highly increase the

TABLE 3.14
Summary of the corrosion rates for Mg–Zn binary alloys.*

Zn content (wt. %)	Additional elements	Corrosive environment	Immersion time	Corrosion rate (mm/year)	Procedure	Ref.
1	—	Hank's, 37°C	25 days	6.22	H	[15]
0.5	—	SBF, 37°C	20 days	0.99	H	[152]
1	—	SBF, 37°C	20 days	1.17	H	[152]
1.5	—	SBF, 37°C	20 days	1.32	H	[152]
2	—	SBF, 37°C	20 days	1.32	H	[152]
0.5 (backward extruded)	—	SBF, 37°C	20 days	0.52	H	[152]
1 (backward extruded)	—	SBF, 37°C	20 days	0.54	H	[152]
1.5 (backward extruded)	—	SBF, 37°C	20 days	0.54	H	[152]
2 (backward extruded)	—	SBF, 37°C	20 days	0.63	H	[152]
5	—	Hank's, 37°C	7 days	4.54	WL	[154]
5	0.2 wt. % Sr	Hank's, 37°C	7 days	3.44	WL	[154]
5	0.6 wt. % Sr	Hank's, 37°C	7 days	3.94	WL	[154]
5	1 wt. % Sr	Hank's, 37°C	7 days	3.83	WL	[154]
5	—	0.9% NaCl, 37°C	7 days	6.74	WL	[154]
5	0.2 wt. % Sr	0.9% NaCl, 37°C	7 days	5.92	WL	[154]
5	0.6 wt. % Sr	0.9% NaCl, 37°C	7 days	6.12	WL	[154]
5	1 wt. % Sr	0.9% NaCl, 37°C	7 days	5.95	WL	[154]
6 (extruded)	—	SBF, 37°C	14 days	14.6	WL	[23]
1	—	SBF, 37°C	10 days	22.7	H	[24]
1 (rolled)	—	SBF, 37°C	10 days	20.7	H	[24]
1	—	Hank's, 37°C	10 days	7.7	H	[24]
1 (rolled)	—	Hank's, 37°C	10 days	7.6	H	[24]
6 (extruded)	—	SBF, 37°C	3 days	0.2	WL	[161]
6 (extruded)	—	SBF, 37°C	30 days	0.07	WL	[161]
6 (extruded)	—	SBF, 37°C	3 days	0.2	WL	[145]
6 (extruded)	—	SBF, 37°C	30 days	0.07	WL	[145]
6 (extruded)	—	In vivo, rabbits	14 weeks	2.32	WL	[145]

* All the Mg–Zn samples are cast, unless otherwise specified. Additional information can be found on www.routledge.com/9780367429454.

number of adhered platelets. Finally, Cipriano et al. [170] compared the adhesion density of HUVEC cells on Mg–4Zn–xSr alloys (x = 0.15, 0.5, 1.0, 1.5 wt.%) with that on pure Mg and AZ31 alloys, using either direct or indirect cell culture studies, and the results are gathered in Table 3.15.

Considering the *in vivo* studies, Mg–6Zn alloy rods have been implanted into the femoral shaft of rabbits and gradually absorbed *in vivo* at about 2.32 mm/year degradation rate with newly formed bone surrounding the implant, facilitated by the presence of hydroxyapatite (HA) and other Mg/Ca phosphates as corrosion products [145]. The viscera histology examination (containing heart, liver, kidney and spleen tissues) and the biochemical measurements (including serum magnesium, serum creatinine, blood urea nitrogen, glutamic- pyruvic transaminase and creatine kinase) proved that the

TABLE 3.15

HUVEC adhesion density results for different Mg–Zn–Sr alloys and their comparison with pure Mg and AZ31 alloy. For two different culture times, either direct or indirect cell cultures are used

Alloy	Adhesion density (cells/cm^2)	Culture time (h)	Testing procedure
Mg–4Zn–0.15Sr	2065	4	Indirect culture
Mg–4Zn–0.5Sr	2624	4	Indirect culture
Mg–4Zn–1Sr	3290	4	Indirect culture
Mg–4Zn–1.5Sr	3226	4	Indirect culture
Pure Mg	3161	4	Indirect culture
AZ31	2989	4	Indirect culture
Mg–4Zn–0.15Sr	2194	24	Indirect culture
Mg–4Zn–0.5Sr	2129	24	Indirect culture
Mg–4Zn–1Sr	3290	24	Indirect culture
Mg–4Zn–1.5Sr	3656	24	Indirect culture
Pure Mg	3182	24	Indirect culture
AZ31	2473	24	Indirect culture
Mg–4Zn–0.15Sr	1943	4	Direct culture
Mg–4Zn–0.5Sr	1692	4	Direct culture
Mg–4Zn–1Sr	1587	4	Direct culture
Mg–4Zn–1.5Sr	1629	4	Direct culture
Pure Mg	1963	4	Direct culture
AZ31	1065	4	Direct culture
Mg–4Zn–0.15Sr	292	24	Direct culture
Mg–4Zn–0.5Sr	334	24	Direct culture
Mg–4Zn–1Sr	376	24	Direct culture
Mg–4Zn–1.5Sr	313	24	Direct culture
Pure Mg	584	24	Direct culture
AZ31	522	24	Direct culture

degradation of Mg–Zn did not harm the important organs. These results agrees with those obtained by He et al. [171] that implanted a Mg–5.6Zn rod into a distal femur marrow cavity of the New Zealand rabbit for 14 weeks with no measurable effect on serum magnesium, or on liver or kidney function tests. Similar results were also found in ref. [23], where the authors implanted Mg–6Zn and pure Mg into the bladders of Wistar rats. Tissues stained with hematoxylin and eosin (HE) suggested that both pure Mg and Mg–6Zn alloy exhibited good histocompatibility in the bladder indwelling implantation, and no differences between pure Mg and Mg–6Zn groups were found in bladder, liver and kidney tissues during the 2 weeks implantation. However, Mg–6Zn alloy has been found to be characterized by a higher *in vivo* corrosion rate, that is, 12.6 mm/year compared to 5.7 mm/year for pure Mg. In addition, although it is unlikely that exposure will reach these levels with the use of an Mg–Zn alloy implant, cytotoxicity has been identified *in vitro* with exposure of cells to Zn concentrations higher than 1.25×10^{-4} mol/l Zn^{+2} [148,172].

3.4.2 MG–ZN–ZR ALLOYS

Mg–Zn–Zr alloys, labeled as ZK alloys, have started to be investigated to increase the mechanical properties of Mg–Zn binary alloys, avoiding the costs associated with long solutionizing and aging treatment. Zr, in fact, has been reported to have a highly grain refining ability [173]. In addition, Zr is already used in a wide range of medical implants including dental alloys and relatively inert orthopedic implants, and is widely accepted as biocompatible [174,175]. Emsley, in fact, reported a daily intake of 50 μg Zr to be permissible [176]. Finally, with its strong chemical affinity with oxygen, Zr can hinder the charge transfer process during corrosion by the formation of a ZrO_2 layer [177]. However, excess Zr may increase corrosion because of its precipitation in the matrix and thus the Zr concentration is normally less than 1 wt.%.

3.4.2.1 Mg–Zn–Zr Alloys: Mechanical Properties

Zr is usually used as a grain refiner in magnesium alloys without aluminum, thereby contributing to the strength of these alloys. For instance, the mechanical properties of as-cast Mg–2Zn have been reported to highly increase by adding 0.2 wt.% Zr. The tensile strength, in fact, increases from 145.9 to 186.9 MPa, while the elongation to failure from 12.2% to 18% [178]. Several studies have then been carried out assessing the effects of manufacturing processes on the mechanical properties. Volkova [179], for example, investigated the effect of hydroextrusion and different heat treatments on ZK60, reporting the former to increase the yield strength and the UTS of 47% and 19%, respectively, whereas the elongation to failure decreased from 11.0% to 6.0%. Concerning the heat treatments, a different behavior has been reported depending on whether the material was previously subjected to hydroextrusion. While hydroextruded specimens subjected to quenching and aging were characterized by lower elongation to failure compared to just-aged hydroextruded samples, in the absence of hydroextrusion, the quenching and aging treatment was reported to increase the elongation to failure. In addition, some researchers sought the determination of optimal process parameters. Nair et al. [180] investigated the effects of RAM speed and extrusion temperature on microstructure and mechanical properties of ZK30 alloy, and reported that increasing the extrusion temperature improves the homogeneity of deformation (by increasing nonbasal slip) but a further increase in temperature (>400°C) causes abnormal grain growth (owing to secondary recrystallization). Increase in RAM speed is also found to enhance uniform deformation but further increase in ram speed (>8 mm/s) causes adiabatic heating and flow localization, besides the occurrence of grain growth/secondary recrystallization, which lowers the properties. In conclusion, they reported the best mechanical properties to be obtained with an extrusion temperature of 400°C and a ram speed of 8 mm/s, since uniform deformation can be obtained in the material in these conditions. In the Table 3.16, the mechanical properties of different ZK alloys available in literature are gathered. Manufacturing processes, when different from cast, or heat treatments are reported in brackets in the "Alloy" column.

As it can be seen from Table 3.16, several authors have focused their attention on severe plastic deformation (SPD) methods, and especially on ECAP [181–199,201]. This is because ECAP, and in particular all SPD procedures, improved the mechanical properties of Mg alloys through grain refinement and texture modification. Shear deformation repeatedly imposed at the sharp ECAP die corner, in fact, increases the ductility of Mg alloys by rotating a high portion of basal planes such that they have high Schmid factors for the test loading direction, whereas the grain refinement induced by the continuous dynamic recrystallization associated with dislocation slip [202] increases the strength according to the Hall–Petch relation. The optimization of ECAP process parameters has thus drawn the researchers' attention. Yuan et al. [203], for example, assessed the ECAP die temperature leading to higher mechanical properties. They manufactured two different groups of ZK60 samples using different die temperatures; the first group was obtained by processing the samples at a constant ECAP die temperature of 300°C for 8 passes, while in the second group, the samples were initially processed by ECAP for 4 passes at 250°C and then the processing temperature was decreased to 200°C for the next 2 passes of ECAP and again to 150°C for another 2 passes. They reported the second group of specimens to be characterized by higher yield strength and UTS, that is, 260 and 371 MPa instead of 184 and 342 MPa, respectively, due to the suppression of the dynamic recovery and recrystallization that led to further refined grains. In addition, the reduction in formability was avoided by gradually reducing the temperature.

3.4.2.2 Mg–Zn–Zr Alloys: Corrosion Resistance

The addition of Zn and Zr as alloying element in binary Mg alloys has been reported to reduce the corrosion rate compared to pure Mg [24], besides leading to strength improvement and good cytocompatibility [204]. As mentioned in Section 3.3, Zn has the ability to reduce the corrosion enhancing effects of

TABLE 3.16

Summary of the mechanical properties of Mg–Zn–Zr alloys.*

Alloy	Zr content (wt. %)	Additional element (wt. %)	Yield strength (MPa)	UTS (MPa)	Elongation to failure (%)	Compressive yield strength (MPa)	Ultimate compressive strength (MPa)	Ref.
ZK60 (pressed)	0.58	—	237	313	11.0	—	—	[179]
ZK60 (pressed + aged)	0.58	—	258	326	10.0	—	—	[179]
ZK60 (pressed + quenched + aged)	0.58	—	281	332	11.0	—	—	[179]
ZK60 (HE)	0.58	—	357	395	6.0	—	—	[179]
ZK60 (HE + aged)	0.58	—	272	322	21.0	—	—	[179]
ZK60 (HE + quenched + aged)	0.58	—	280	328	14.5	—	—	[179]
ZK60 (quenched + HE + aged)	0.58	—	323	378	4.5	—	—	[179]
ZK30 (extruded)	0.5	—	232.4	286.8	13.8	—	—	[180]
ZK60 (twin-rolled cast + hot rolled)	0.5	—	425	477	4.1	—	—	[181]
ZK60 (twin-rolled cast + annealed + hot rolled)	0.5	—	407	474	8.0	—	—	[181]
ZK60 (twin-rolled cast + hot rolled)	0.5	—	389	455	7.6	—	—	[181]
ZK60 (twin-rolled cast + annealed + hot rolled)	0.5	—	374	435	13.1	—	—	[181]
ZK60 (solution treated + HRSDR)	0.53	—	—	351.1	15.0	—	—	[182]
ZK60 (HRSDR)	0.53	—	—	345.3	13.9	—	—	[182]
ZK60 (solution treated + HRSDR + annealed)	0.53	—	—	298.9	19.8	—	—	[182]
ZK60 (HRSDR + annealed)	0.53	—	—	297.9	30.8	—	—	[182]
ZK60 (HSRR)	0.6	—	223	311	18.3	—	—	[183]
ZK60 (HSRR)	0.6	0.2 Gd	227	307	25.3	—	—	[183]
ZK60 (HSRR)	0.6	0.5 Gd	235	318	23.2	—	—	[183]
ZK60 (HSRR)	0.6	0.8 Gd	242	327	22.0	—	—	[183]

(Continued)

TABLE 3.16 (Cont.)

Alloy	Zr content (wt. %)	Additional element (wt. %)	Yield strength (MPa)	UTS (MPa)	Elongation to failure (%)	Compressive yield strength (MPa)	Ultimate compressive strength (MPa)	Ref.
ZK60 (extruded)	0.5	—	230.1	280.9	17.5		—	[184]
ZK60 (extruded + ECAP)	0.5	—	231.5	291.2	27.2		—	[184]
ZK60 (extruded)	0.48	—	293.3	337.2	15.7		—	[185]
ZK60 (extruded + ECAP)	0.48	—	235.6	314.1	17.2		—	[185]
ZK60 (extruded + ECAP + annealed)	0.48	—	120.1	271.4	42.0		—	[185]
ZK60 (ECAP)	0.56	—	199.5	349.3	25.2		—	[186]
ZK60 (rolled)	0.56	—	332.4	351.8	5.5		—	[186]
ZK60 (ECAP + rolled)	0.56	—	353.0	388.0	7.2		—	[186]
ZK60 (ECAP + annealed + rolled)	0.56	—	395.3	429.1	9.2		—	[186]
ZK60 (extruded)	0.5	—	222	264	7.4		—	[187]
ZK60 (extruded + ECAP)	0.5	—	310	351	17.1	—	—	[187]
ZK60	0.5	—	150	213	6	120	—	[141]
ZK60 (extruded)	0.5	—	290	335	16	250	—	[141]
ZK60 (extruded)	0.5	—	263	318	10.4	—	—	[188]
ZK60 (extruded + ECAP)	0.5	—	183	291	9.6		—	[188]
ZK60 (extruded + ECAP + extruded)	0.5	—	313	377	8.2		—	[188]
ZK60 (extruded + ECAP + annealed + extruded)	0.5	—	328	344	13.8		—	[188]
ZK60 (MAF)	0.36	—	202	315	13		—	[189]
ZK60 (CEC)	0.5	—	215	289	37.8		—	[190]
ZK60 (extruded)	0.5	—	248	316	13.9		—	[190]
ZK60 (ECAP)	0.6	—	175	266	25.9		—	[191]

ZK21 (LSRMF)	0.45	—	161	260.3	18.3	—	—	[192]
ZK21 (HSRMF)	0.45	—	230.7	407.8	9.6	—	—	[192]
ZK60 (extruded)	0.6	—	166	250	18.5	—	—	[193]
ZK60 (extruded + T4 + ECAP)	0.6	—	175	266	31.9	—	—	[193]
ZK60 (extruded)	—	—	262.3	325.2	21.0	—	—	[194]
ZK60 (extruded)	—	0.5 Er	290.8	340.6	20.3	—	—	[194]
ZK60 (extruded)	—	1 Er	273.8	332.7	18.4	—	—	[194]
ZK60 (extruded)	—	2 Er	281.3	339.7	19.0	—	—	[194]
ZK60 (extruded)	—	4 Er	271.3	333.7	20.4	—	—	[194]
ZK60 (extruded + solid solution)	—	2 Er	212	305	16	—	—	[194]
ZK60 (extruded + solid solution + aged)	—	2 Er	247	315	19	—	—	[194]
ZK60	0.5	—	135.8	170.7	3.4	—	—	[195]
ZK60 (SSTT)	0.5	—	145.5	216.6	4.0	—	—	[195]
ZK60 (RAP)	0.5	—	210.1	302.6	12.1	—	—	[195]
ZK60	—	—	—	208.5	16.0	—	—	[196]
ZK60 (rolled)	—	—	—	349.6	5.0	—	—	[196]
ZK60 (rolled + EPT)	—	—	—	307.7	32.2	—	—	[196]
ZK60 (hot pressed)	0.65	—	200	310	17	—	—	[197]
ZK60 (hot pressed + MIF)	0.65	—	145	280	37	—	—	[197]
ZK60 (hot pressed + IR)	0.65	—	180	320	24	—	—	[197]
ZK60 (ECAP)	0.56	—	204.1	355.9	25.1	—	—	[198]
ZK60 (rolled)	0.56	—	336.2	357.0	5.6	—	—	[198]
ZK60 (ECAP + rolled)	0.56	—	397.4	430.1	9.4	—	—	[198]
ZK31 (extruded)	0.8	—	227.5	306.1	20	—	—	[168]
ZK31 (extruded)	0.8	0.3 Sr	322.4	376.4	16	—	—	[168]

(Continued)

TABLE 3.16 (Cont.)

Alloy	Zr content (wt. %)	Additional element (wt. %)	Yield strength (MPa)	UTS (MPa)	Elongation to failure (%)	Compressive yield strength (MPa)	Ultimate compressive strength (MPa)	Ref.
ZK60 (extruded)	0.57	—	—	312.5	18.0	—	—	[199]
ZK60 (extruded + HRDSR)	0.57	—	—	408.5	12.3	—	—	[199]
ZK60 (extruded + HRDSR + annealed)	0.57	—	—	367.9	15.8	—	—	[199]
ZK60 (DCC)	0.64	—	105.9	234.8	105.9	—	—	[200]
ZK60 (DCC + VFUT)	0.64	—	122.9	279.0	122.9	—	—	[200]
ZK60 (extruded)	0.55	—	216.0	285.4	14.1	—	—	[201]
ZK60 (extruded + ECAP)	0.55	—	142.2	309.0	30.0	—	—	[201]

* All the Mg–Zn–Zr samples are cast, unless otherwise specified. Additional information can be found on www.routledge.com/9780367429454.

all of the common impurities including iron, nickel and copper. In addition, Zn has been reported to increase the corrosion resistance due to its high solid solubility limit [158]. On the other hand, Zr, besides hindering the charge transfer process during corrosion by the formation of a Zr oxide layer [177], reduces the adverse effects of iron contamination on the corrosion resistance of Mg alloys [205]. However, excess Zr may increase corrosion because of precipitation of Zr in the matrix and thus the Zr concentration is normally less than 1 wt.%. However, when alloyed together they lead to different results. While ZK60 has been reported to be characterized by a higher corrosion rate than pure Mg [206–208], ZK30 and ZK21 have been found to corrode slower than pure Mg [209,210]. This is due to the formation of second-phase particles (Mg_7Zn_3) and microgalvanic cells. It has been reported that a Zn addition limited to 1–3% improves the corrosion resistance [211], while, when the Zn content goes beyond 3 wt.%, a considerable amount of the Mg_7Zn_3 phase will form in the Mg matrix resulting in microgalvanic coupling with the matrix, thus accelerating the corrosion of the material. However, the homogeneous distribution of these second phases would act as a corrosion barrier. To this aim, SPD techniques are exploited to produce ultra-fine grained materials with sub-micrometer grain size and a homogeneous distribution of the nano-sized second phases. In fact, Mostaed et al. reported an improved corrosion resistance of ultra-fine grained ZK60 alloy compared to the extruded one due to second-phases redistribution [212], and further data about polarization curves and corrosion rate are reported in Tables 3.17 and 3.18, respectively.

3.4.2.3   Mg–Zn–Zr Alloys: Biocompatibility

When dealing with studying new biocompatible material, the ternary alloy Mg–Zn–Zr has been believed to be highly suitable for biomedical applications. Zn, in fact, is a trace element, involved in a wide range of physiological functions, and Zr has already been used in a wide range of medical implants due to its known biocompatibility (its daily intake is 4.15 mg [213]; however, higher concentrations may lead to lung cancer, liver cancer, nasopharyngeal cancer and breast cancer [15]). This believed biocompatibility has then been confirmed by *in vitro* and *in vivo* tests. Zhang et al., in fact, explored the feasibility of magnesium alloys for bone repair, assessing a ZK60 alloy with a reduced corrosion rate as a bone graft substitute [214]. The alloy was used to repair the artificial bone defects at the lateral tibial plateau in minipigs, and the results were compared with those obtained using calcium sulfate pellet, a conventional substitute clinically used to repair bone defects [215,216]. The defect repaired by the magnesium alloy displayed a bony morphology similar to that of the normal bone, and the average bone healing rate of the magnesium alloy defects was higher than that of the calcium sulphate pellet-treated ones over the first 4 months. In addition, no signs of local inflammation, infection or tissue reaction were observed, which is in agreement with Lin et al. [217]. They implanted ZK60 rods into both rabbit femur bones. They reported bone to directly grow on the implant surface and in the degraded cavities, indicating the excellent osteo-compatibility, and the biological

TABLE 3.17

Summary of the DC polarization curve results for Mg–Zn–Zr ternary alloys.*

Alloy	Zr content (wt.%)	Additional elements	Corrosive environment	E_{corr} (V)	i_{corr} ($\mu A/cm^2$)	Ref.
ZK60	0.5	—	AU, 37°C	−1.56	41.9	[207]
ZK60	0.55	—	SBF, 37°C	−1.39	101.4	[208]
ZK60	0.55	0.9 wt.% Y	SBF, 37°C	−1.43	44.2	[208]
ZK60 (extruded + solution treated + HRDSR)	0.53	—	Hank's, 37°C	−1.42	13.8	[182]
ZK60 (extruded + HRDSR)	0.53	—	Hank's, 37°C	−1.40	10.6	[182]
ZK60 (extruded + solution treated + HRDSR + annealed)	0.53	—	Hank's, 37°C	−1.43	16.9	[182]
ZK60 (extruded + HRDSR + annealed)	0.53	—	Hank's, 37°C	−1.40	9.2	[182]
ZK60	0.5	—	SBF, 37°C	−1.65	409	[177]
ZK60	0.5	—	DMEM, 37.4°C	−1.49	37.1	[206]
ZK60 (T4)	0.5	—	DMEM, 37.4°C	−1.55	38.1	[206]
ZK21	0.6	—	Hank's, 37°C	−1.57	—	[210]
ZK21	0.6	0.2 wt.% Sc	Hank's, 37°C	−1.55	—	[210]
ZK60 (extruded)	0.48	—	PBS, 37°C	−1.49	60.4	[212]
ZK60 (extruded + ECAP)	0.48	—	PBS, 37°C	−1.46	56.2	[212]

* All the Mg–Zn–Zr samples are cast, unless otherwise specified.

functionality of the surrounding tissues also was not significantly affected during the entire serving period of the implant. However, ZK60 alloy degraded too fast in the transcortical model to meet the clinical requirement. The results agreed with those reported in refs. [218,219]. Good cytocompatibility has been also found *in vitro* when the corrosion rate is low [220–222]. Huan et al. [209], in fact, reported the ZK60 alloy to be cytotoxic due to its high hydrogen evolution. In addition, they assessed the ZK30 cytocompatibility carrying out indirect contact tests on rabbit bone marrow stromal cells (rBMSCs) and they used biocompatible HA as reference. They reported ZK30 to display a cytocompatibility quite similar to HA after 1-day incubation. However, after a prolonged incubation period of 7 days, the rBMSC proliferation in the extract of ZK30 sample was found to be significantly higher than that of HA sample. They attributed the enhanced cell proliferation of ZK30 to the moderate Mg ion release that, due to its functional roles and presence in bone tissue, may have stimulatory effects on the growth of new bone tissue [223], in agreement with Zreiqat et al. [224] who reported

TABLE 3.18
Summary of the corrosion rates for Mg–Zn–Zr ternary alloys.*

Alloy	Zr content (wt.%)	Additional elements (wt.%)	Corrosive environment	Immersion time	Corrosion rate (mm/ year)	Procedure	Ref.
ZK60	0.5	—	AU, 37°C	28 days	1.0	WL	[207]
ZK60	0.55	—	SBF, 37°C	10 days	2.21	WL	[208]
ZK60	0.55	0.9 wt.% Y	SBF, 37°C	10 days	1.21	WL	[208]
ZK60 (extruded + solution treated + HRDSR)	0.53	—	Hank's, 37°C	1 day	2.65	WL	[182]
ZK60 (extruded + HRDSR)	0.53	—	Hank's, 37°C	1 day	2.47	WL	[182]
ZK60 (extruded + solution treated + HRDSR + annealed)	0.53	—	Hank's, 37°C	1 day	2.92	WL	[182]
ZK60 (extruded + HRDSR + annealed)	0.53	—	Hank's, 37°C	1 day	1.34	WL	[182]
ZK60 (extruded + solution treated + HRDSR + annealed)	0.53	—	Hank's, 37°C	1 day	1.56	H	[182]
ZK60 (extruded + HRDSR + annealed)	0.53	—	Hank's, 37°C	1 day	0.76	H	[182]
ZK60	0.5	—	DMEM, 37.4°C	7 days	0.4	WL	[206]
ZK60	0.5	—	DMEM, 37.4°C	14 days	1.1	WL	[206]
ZK60	0.5	—	DMEM, 37.4°C	21 days	1.5	WL	[206]
ZK60 (T4)	0.5	—	DMEM, 37.4°C	7 days	0.5	WL	[206]
ZK60 (T4)	0.5	—	DMEM, 37.4°C	14 days	0.46	WL	[206]
ZK60 (T4)	0.5	—	DMEM, 37.4°C	21 days	0.9	WL	[206]

(Continued)

TABLE 3.18 (Cont.)

Alloy	Zr content (wt.%)	Additional elements (wt.%)	Corrosive environment	Immersion time	Corrosion rate (mm/ year)	Procedure	Ref.
ZK60	0.6	—	Hank's, 37°C	12 weeks	16	WL	[209]
ZK30	0.6	—	Hank's, 37°C	21 weeks	0.09	WL	[209]
ZK21	0.6	—	Hank's, 37°C	7 days	0.06	H	[210]
ZK21	0.6	0.2 wt.% Sc	Hank's, 37°C	7 days	0.10	H	[210]
ZK21	0.6	—	Hank's, 37°C	7 days	0.34	WL	[210]
ZK21	0.6	0.2 wt.% Sc	Hank's, 37°C	7 days	0.327	WL	[210]
ZK60 (extruded)	0.48	—	PBS, 37°C	7 days	4.11	WL	[212]
ZK60 (extruded + ECAP)	0.48	—	PBS, 37°C	7 days	2.82	WL	[212]
ZK31 (extruded)	0.8	—	SBF, 37°C	15 days	4.1	WL	[168]

* All the Mg–Zn–Zr samples are cast, unless otherwise specified. Additional information can be found on www.routledge.com/9780367429454.

magnesium ion modified bioceramic substrates to enhance the human bone-derived cell adhesion. However, exposure to the large concentrations of Mg^{2+}, especially over 200 µg/ml, results in an adverse effect on cell growth, as shown by Jin et al. [225] that, carrying out indirect tests on MC3T3-E1 cells, reported ZK60 to be cytotoxic due to its high corrosion rate. Further data on the biocompatibility of Mg–Zn–Zr alloys, in particular in terms of hemolysis and cell viability, are gathered in Table 3.19, reporting no cytotoxic effect of Mg–Zn–Zr alloys characterized by a low corrosion rate.

3.4.3 MG–ZN–CA ALLOYS

Ca and Zn are the most ideal elements to be used as bone substitute materials. The advantages of Zn have already been reported in Section 3.3. Concerning Ca, it represents the main composition of human bone and therefore it can accelerate bone healing [226,227]. In addition, the introduction of Ca has been reported to refine the microstructure of the Mg–Zn binary alloys during solidification, extrusion and rolling [228] and to improve the age hardenability [229]. Although the main demerits as implant materials of cast Mg–Zn–Ca (ZX) alloys consist of their poor corrosion resistance and mechanical properties that are not beneficial to their clinical applications, they can be improved by heat treatment and extrusion [230]. MTT results also clearly indicated that

TABLE 3.19

Summary of the biocompatibility of Mg–Zn–Zr alloys.*

Alloy	Zr content (wt.%)	Test environment	Cell viability			Hemolysis (%)	Ref.
			Cell type, procedure	Time of culture	Result (%)		
ZK60	0.52	DMEM, 37°C	MC3T3-E1, IC	1 day	27.6	—	[225]
ZK60	0.52	DMEM, 37°C	MC3T3-E1, IC	3 days	20.7	—	[225]
ZK60	0.52	DMEM, 37°C	MC3T3-E1, IC	6 days	3.1	—	[225]
ZK40	0.5	DMEM, 37°C	MC3T3-E1, IC	3 days	75.5	—	[206]
ZK40 (T4)	0.5	DMEM, 37°C	MC3T3-E1, IC	3 days	53.0	—	[206]
ZK30 (homogenized + extruded)	0.8	RPMI, 37°C	L-929, IC	1 day	121	—	[168]
ZK30 (homogenized + extruded)	0.8	RPMI, 37°C	L-929, IC	3 days	123	—	[168]
ZK30 (homogenized + extruded)	0.8	RPMI, 37°C	L-929, IC	5 days	131	—	[168]

* All the Mg–Zn–Zr samples are cast, unless otherwise specified. Additional information can be found on www.routledge.com/9780367429454.

Ca did not induce cytotoxicity [231]. In addition, Mg–Zn–Ca BMGs are being studied as an alternative biodegradable fixation material to Mg–Zn–Ca alloys and they will be treated in a specific section.

3.4.3.1 Mg–Zn–Ca Alloys: Mechanical Properties

As mentioned in the Section 3.4.3., the addition of calcium leads to microstructure refinement and to an age hardening effect. However, the Ca content that is suggested to be selected should be less than 1 wt.% (the solubility limit of Ca in Mg) to avoid excess formation of Mg_2Ca phase that hinder the age hardening effect and does not result in further grain refinement [232,233]. In addition, the Zn/Ca atomic ratio is recommended to be within the range mentioned in literature (1.2–2.0) for optimum age hardening effect and corrosion resistance [233–235]. Nevertheless, the main advantage of Ca addition on Mg–Zn alloys relies on the increased ductility of the base alloy after manufacturing processes involving plastic deformation [236–250]. Du et al. carried out tensile tests on either extruded Mg–3Zn or extruded Mg–3Zn–0.05Ca, reporting the latter to be characterized by a comparable tensile strength but a higher elongation to failure (28.2% instead of 25.3%) [251]. The mechanical properties of over studied Mg–Zn–Ca systems are reported in Table 3.20, confirming a high

TABLE 3.20

Summary of Mg–Zn–Ca alloys mechanical properties.*

Zn content (wt.%)	Ca content (wt.%)	Additional element (wt.%)	Yield strength (MPa)	UTS (MPa)	Elongation to failure (%)	Compressive yield strength (MPa)	Ultimate compressive strength (MPa)	Ref.
6 (homogenized + extruded)	0.2	0.4 Ag	153	283.4	25	103	430	[238,239]
6 (homogenized + extruded)	0.2	0.4 Ag, 0.6 Zr	289	353.6	17	246	550	[238,239]
6 (homogenized + extruded + T6)	0.2	0.4 Ag, 0.6 Zr	325	355	14	—	—	[238,239]
3.1 (homogenized + extruded)	0.05	—	128.5	244.3	28.2	—	—	[251]
0.9 (annealed + extruded)	0.25	—	—	—	—	114	—	[244]
0.9 (annealed + extruded + pre-compressed)	0.25	—	—	—	—	148	—	[244]
0.9 (annealed + extruded + pre-compressed + aged)	0.25	—	—	—	—	173	—	[244]
0.21 (homogenized)	0.3	0.14 Mn	36	134	7.3	33.1	—	[245]
0.21 (homogenized + extruded)	0.3	0.14 Mn	165.2	221.1	30.1	127.2	—	[245]
6 (homogenized + extruded)	0.4	—	169	276	21.4	—	—	[236]
6 (homogenized + extruded)	0.8	—	230	304	15.3	—	—	[236]
5.3 (homogenized + extruded)	0.5	0.5 Ce/La	164.2	258.9	21.1	—	—	[246]
5.3 (homogenized + extruded + rolled)	0.5	0.5 Ce/La	316	338	9.5	—	—	[246]
6 (indirect extruded)	0.2	—	148	275	26	135	443	[247]
6 (indirect extruded)	0.2	0.8 Zr	310	357	18	261	523	[247]
1 (extruded)	0.2	—	—	239.8	39.0	—	—	[228]
1 (extruded)	0.5	—	105	210	44	—	—	[228]
0.3 (homogenized + rolled + annealed)	0.1	—	92.5	186	24	—	—	[252]

2 (homogenized + rolled)	0.8	—	99	175	7	—	—	[248]
2 (homogenized + rolled)	0.3	—	80	210	21	—	—	[248]
1 (homogenized + rolled)	0.3	—	107	210	24	—	—	[248]
1 (homogenized + rolled)	0.1	—	72	200	37	—	—	[248]
0.5 (extruded)	0.2	0.2 Ce	83.9	273.8	28.1	68.4	—	[249]
1 (extruded)	0.2	0.2 Ce	111.4	323.8	37.3	90.7	—	[249]
1.5 (extruded)	0.2	0.2 Ce	131.0	336.4	42.4	111.7	—	[249]
2 (extruded)	0.2	0.2 Ce	183.1	345.2	29	137	—	[249]
0.21 (homogenized + extruded)	0.30	0.14 Mn	—	193.0	31.1	—	—	[250]
0.53 (homogenized + extruded)	0.24	0.27 Mn	—	204.9	32.1	—	—	[250]
0.71 (homogenized + extruded)	0.36	0.07 Mn	108	220	37	—	—	[250]
6.2 (homogenized + extruded)	0.7	—	205	286.2	18.4	—	—	[237]
6.2 (homogenized + extruded)	0.7	0.22 Ce	285	325.3	16.1	—	—	[237]
5.25 (extruded)	0.6	—	178	279	25.9	—	—	[240]
5.25 (extruded + ECAP)	0.6	—	246	332	15.5	—	—	[240]
5.12 (solution treated + extruded)	0.32	—	—	314.3	13.2	—	—	[241]
5.12 (solution treated + extruded + ECAP)	0.32	—	—	296.1	18.2	—	—	[241]
3 (solution treated + quenched + RTE)	0.2	—	171.6	241.7	21.2	—	—	[242]
3 (solution treated + quenched + extruded)	0.2	—	145.5	240.1	36.7	—	—	[242]
5.25 (extruded + ECAP)	0.6	—	—	—	—	200	288.7	[243]
5.25 (extruded + ECAP + annealed)	0.6	—	—	—	—	136	412	[243]
2 (homogenized + extruded)	0.5	—	127	212	34.3	—	—	[254]
5.3	0.2	0.5 Ce	71	169	6.6	—	—	[253]
5.3 (homogenized + extruded)	0.2	0.5 Ce	268	320	14.7	—	—	[253]

* All the Mg–Zn–Ca samples are cast, unless otherwise specified. Additional information can be found on www.routledge.com/9780367429454.

elongation to failure in alloys subjected to plastic deformation processes, that is, extrusion, rolling, ECAP. Mg–Zn–Ca alloys have in fact been reported to develop a weaker basal texture during the dynamic recrystallization induced by the plastic deformation. Zeng et al. [252] reported the texture intensity of hot-rolled and annealed Mg–0.4Zn alloy to be 12 mrd, whereas 3.7 mrd for hot-rolled and annealed Mg–0.3Zn–0.1Ca (Figure 3.9).

Two probable mechanisms have been proposed; one is the particle-stimulated nucleation mechanism caused by the Ca-containing second phase [228,253] and the other is the reduction of c/a ratio of the lattice caused by the solute Ca atoms in Mg matrix [248,254].

3.4.3.2 Mg–Zn–Ca Alloys: Corrosion Resistance

According to the phase diagram of binary Mg–Ca, the maximum solubility of Ca in Mg was reported as 0.2 wt.% at room temperature in the equilibrium state [255]. Zhang et al. [159,256] found that 0.2 wt.% Ca addition reduces the degradation rate of as-cast Mg–4Zn alloy to approximately one-third due to a finer microstructure, whereas the degradation rate of as-cast Mg–4Zn–xCa alloy increases with increasing Ca content (x = 0.2–2 wt.%) due to the formation of coarse precipitates, such as Mg_2Ca, $Ca_2Mg_6Zn_3$ and $Ca_2Mg_5Zn_{13}$. The percentage of such compounds for the Mg–Zn–Ca ternary alloys increases by increasing the content of Zn and Ca. In particular, Bakhsheshi-Rad et al.

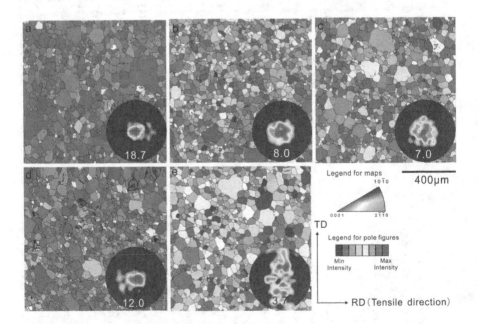

FIGURE 3.9 EBSD orientation maps and corresponding basal pole figures showing microstructure and textures of hot-rolled and subsequently annealed Mg–0.4Zn (a) and Mg–0.3Zn–0.1Ca (b). Reprinted with permission from Elsevier [252].

[257] pointed out that very high Zn content is detrimental because of the formation of high amounts of $Ca_2Mg_6Zn_3$, which is reported to have a more rapid corrosion rate. Consequently, reduction of the Zn content might be an effective recipe for achieving better biocorrosion resistance. Such a simple procedure, however, would result in significant loss in strength because of reduced solid solution hardening. Hänzi et al. [258] reported that the solid solution hardening of Zn can easily be calculated to be roughly 20 MPa/wt.%. Hence, further measures have to be taken to compensate the reduced solid solution hardening. According to refs. [259,260], the Hall–Petch coefficient for fine-grained Mg alloys is about 5 MPa mm$^{1/2}$. Accordingly, a reduction in grain size is aimed to compensate the loss in mechanical properties necessary to improve the corrosion resistance, and thus SPD processes have been widely investigated (see Table 3.21 on the mechanical properties). In addition, grain refinement is reported to be beneficial also for the corrosion resistance. Besides reducing the corrosion rate due to a higher number of grain boundaries that act as physical barrier [261], Zhang et al. [262] reported a Mg–Zn–Ca alloy processed by high-pressure torsion to be characterized by a slower and uniform corrosion compared to the alloy in the as-received condition due to finer and more uniformly distributed second particles [262], agreeing with Gao et al. [263]. In Tables 3.21 and 3.22, a summary of the corrosion behavior of Mg–Zn–Ca alloys in terms of polarization curves (Table 3.21) and in terms of corrosion rate (Table 3.22) is reported.

3.4.3.3 Mg–Zn–Ca Alloys: Biocompatibility

The ternary Mg–Zn–Ca alloy has been developed in the biomedical field aiming to improve the mechanical properties of the binary alloy Mg–Ca through the

TABLE 3.21

Summary of the DC polarization curve results for Mg–Zn–Ca ternary alloys.*

Zn content (wt.%)	Ca content (wt.%)	Additional elements	Corrosive environment	E_{corr} (V)	i_{corr} (μA/cm^2)	Ref.
2 (homogenized + aged)	1	—	DMEM, 37°C	−1.58	195.3	[164]
2 (homogenized + aged)	2	—	DMEM, 37°C	−1.59	217.7	[164]
2 (homogenized + aged)	3	—	DMEM, 37°C	−1.61	252.4	[164]
2	0.24	—	SBF, 37°C	−1.71	530	[263]
2 (HPT)	0.24	—	SBF, 37°C	−1.76	3.3	[263]
2	0.24	—	SBF, 37°C	−1.75	171.9	[262]
2 (homogenized + extruded + HPT)	0.24	—	SBF, 37°C	−1.73	21.9	[262]
2	2	0.5 Mn	SBF, 37°C	−1.62	78.3	[235]
4	2	0.5 Mn	SBF, 37°C	−1.65	99.6	[235]
7	2	0.5 Mn	SBF, 37°C	−1.73	174.1	[235]

* All the Mg–Zn–Ca samples are cast, unless otherwise specified.

TABLE 3.22

Summary of the corrosion rates for Mg–Zn–Ca ternary alloys.*

Zn content (wt.%)	Ca content (wt.%)	Additional elements (wt.%)	Corrosive environment	Immersion time	Corrosion rate (mm/ year)	Procedure	Ref.
2 (homogen-ized + aged)	1	—	DMEM, 37°C	6 weeks	0.41	WL	[164]
2 (homogen-ized + aged)	2	—	DMEM, 37°C	6 weeks	1.38	WL	[164]
2 (homogen-ized + aged)	3	—	DMEM, 37°C	6 weeks	1.52	WL	[164]
1	1	—	Hank's, 37°C	30 days	2.13	WL	[256]
2	1	—	Hank's, 37°C	30 days	2.38	WL	[256]
3	1	—	Hank's, 37°C	30 days	2.92	WL	[256]
4	1	—	Hank's, 37°C	30 days	4.42	WL	[256]
5	1	—	Hank's, 37°C	30 days	6.15	WL	[256]
6	1	—	Hank's, 37°C	30 days	9.21	WL	[256]

* All the Mg–Zn–Ca samples have are cast, unless otherwise specified. Additional information can be found on www.routledge.com/9780367429454.

addition of the Zinc and to improve the biocompatibility of the binary alloy Mg–Zn by means of Ca. Ca is essential for living organisms and is also a major component of human bone. In addition, it forms calcium phosphates (CaP) during the degradation process, which could provide a more suitable local environment for bone mineralization [264]. Cipriano et al. [265] studied the *in vitro* cytocompatibility of Mg–xZn–0.5Ca alloys, and reported that BMSC adhesion density directly on the degradable Mg–xZn–0.5Ca alloys was comparable to the positive controls at 24, 48 and 72 h culture being the ions concentrations well below the cytotoxic values (30 mM, 60 μM and 30 mM for Mg^{2+}, Zn^{2+} and Ca^{2+}, respectively [232]). In addition, the BMSC culture with the Mg–xZn–0.5Ca alloys only reached a maximum pH of 8.1 and thus did not cause adverse effects in BMSC adhesion or morphology. In Table 3.23, other results about biocompatibility of Mg–Zn–Ca alloys are reported, confirming the results obtained by Cipriano et al. [265].

3.4.4 Mg–Zn–Ca BMGs

Mg–Zn–Ca BMGs have been studied as an alternative biodegradable fixation material with superior strength and corrosion resistance in comparison with the traditional cast Mg–Zn–Ca alloys [266,267]. These superior properties are due to their amorphous structure. They are in fact prepared through a rapid cooling method that does not leave enough time to the crystalline phases to nucleate and grow, leading to the freeze of the material in an amorphous

TABLE 3.23

Summary of the biocompatibility of Mg–Zn–Ca alloys.*

Zn content (wt. %)	Ca content (wt.%)	Test environment	Cell viability			Hemolysis (%)	Ref.
			Cell type, procedure	Time of culture	Result (%)		
2 (quenched + aged)	1	DMEM, 37°C	ASCs, IC	1 day	114.3	—	[164]
2 (quenched + aged)	2	DMEM, 37°C	ASCs, IC	1 day	111.4	—	[164]
2 (quenched + aged)	3	DMEM, 37°C	ASCs, IC	1 day	102.0	—	[164]
2 (quenched + aged)	1	DMEM, 37°C	ASCs, IC	3 days	105.3	—	[164]
2 (quenched + aged)	2	DMEM, 37°C	ASCs, IC	3 days	102.5	—	[164]
2 (quenched + aged)	3	DMEM, 37°C	ASCs, IC	3 days	103.3	—	[164]
1	1	DMEM, 37°C	L-929, IC	1 day	86.8	—	[256]
2	1	DMEM, 37°C	L-929, IC	1 day	82.2	—	[256]
3	1	DMEM, 37°C	L-929, IC	1 day	84.8	—	[256]
1	1	DMEM, 37°C	L-929, IC	3 days	74.2	—	[256]
2	1	DMEM, 37°C	L-929, IC	3 days	75.2	—	[256]
3	1	DMEM, 37°C	L-929, IC	3 days	76.2	—	[256]
1	1	DMEM, 37°C	L-929, IC	5 days	81.5	—	[256]
2	1	DMEM, 37°C	L-929, IC	5 days	74.5	—	[256]
3	1	DMEM, 37°C	L-929, IC	5 days	75.2	—	[256]
1	1	DMEM, 37°C	L-929, IC	7 days	74.8	—	[256]
2	1	DMEM, 37°C	L-929, IC	7 days	69.9	—	[256]
3	1	DMEM, 37°C	L-929, IC	7 days	72.2	—	[256]

* All the Mg–Zn–Ca samples are cast, unless otherwise specified. Additional information can be found on www.routledge.com/9780367429454

glassy state that exceeds the solubility limits of alloying elements in the equilibrium state (Figure 3.10 indicates the range of possible compositions with good glass-forming ability [268]).

Since the atoms either have short-range ordering or are completely disordered, metallic glass may not have crystalline defects such as dislocations and grain boundaries, leading to a high tensile strength and good corrosion resistance [269]. However, the main limitations of Mg–Zn–Ca BMGs are the high cost (high-pure elements), limited dimension (based on the literature the largest cross section is 5 mm [270]) and brittle failure without noticeable plastic strain at room temperature (less than 1% [271]).

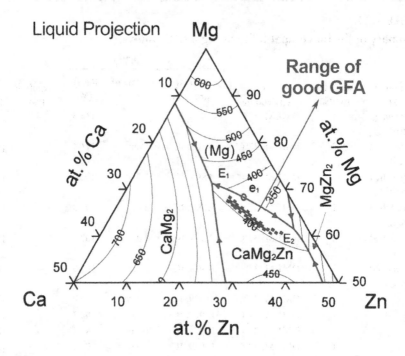

FIGURE 3.10 Range of good glass-forming ability compositions (indicated by the green arrow) superimposed to the Mg–Zn–Ca ternary phase diagram [268].

3.4.4.1 Mg–Zn–Ca BMGs: Mechanical Properties

The BMGs are characterized by higher mechanical properties than their crystalline counterparts. Li et al. [272] studied the compressive strength of a glassy Mg–Zn–Ca alloy, that is, $Mg_{68}Zn_{28}Ca_4$, resulting in a strength far higher than the crystalline Mg–4Zn–0.5Ca, that is, 828 MPa instead of 273 MPa [230]. In addition, they found that the best composition range defined both by strength and plastic deformation should be 3–5% (mole fraction) Ca for $Mg_{72-x}Zn_{28}Ca_x$ (x = 0–6). However, all these Mg–Zn–Ca BMGs fracture in the elastic regime or with a limited plastic deformation (max. 1.3%) due to their inhomogeneous deformation behavior [271], which limits their application. There have hitherto been a number of studies reporting on effective methods that enhance the strength as well as the plasticity of Mg-based BMGs through the dispersion of ductile metallic particles such as those of Mo, Fe, Nb and Ti [7–17]. The compressive elongation to failure of $Mg_{69}Zn_{27}Ca_4$ was increased to 15% by adding 3 wt.% Fe [273]. These ductile metallic particles can absorb the shear strain energy of shear bands, confine the propagation of shear bands and significantly improve their plasticity. Moreover, the addition of ceramic particles, such as Al_2O_3, to the $Mg_{67}Zn_{28}Ca_5$ BMG in a proportion of 1.5 vol.% increases its compressive failure strain from 1.2% to 2.6% [274]. Further results are reported in Table 3.24.

TABLE 3.24
Summary of the mechanical properties of Mg–Zn–Ca BMGs.*

Zn content (at. %)	Ca content (at. %)	Additional element (at. %)	Yield strength (MPa)	UTS (MPa)	Elongation to failure (%)	Compressive yield strength (MPa)	Ultimate compressive strength (MPa)	Ref.
25	5	—	—	—	—	—	565.8	[267]
30	4	—	—	—	—	—	531.2	[267]
27	4	—	—	—	—	—	550	[273]
27	4	3 wt. % Fe	—	—	—	—	648	[273]
28 (wire)	5	—	—	817	—	—	—	[53]
28	1	—	—	—	—	441	441	[272]
28	2	—	—	—	—	442	442	[272]
28	3	—	—	—	—	591	675	[272]
28	4	—	—	—	—	611	828	[272]
28	5	—	—	—	—	622	662	[272]
28	6	—	—	—	—	496	496	[272]
35	3	—	—	—	—	564	564	[272]
33	4	—	—	—	—	582	582	[272]
30	4	—	—	—	—	590	590	[272]
25	5	—	—	—	—	620	642	[272]
20	5	—	—	—	—	420	510	[272]
15	5	—	—	—	—	307	513	[272]
28	5	—	—	—	—	602	650	[274]
28	5	0.66 vol. % Al_2O_3 particles	—	—	—	640	690	[274]
28	5	1.2 vol. % Al_2O_3 particles	—	—	—	710	750	[274]
28	5	1.5 vol. % Al_2O_3 particles	—	—	—	750	780	[274]

* All the Mg–Zn–Ca BMGs are cast, unless otherwise specified. Additional information can be found on www.routledge.com/9780367429454

3.4.4.2   Mg–Zn–Ca BMGs: Corrosion Resistance

Gu et al. [267] compared the corrosion behavior of two metallic glasses, $Mg_{66}Zn_{30}Ca_4$ and $Mg_{70}Zn_{25}Ca_5$, with that of rolled pure Mg, and reported that the BMGs are characterized by lower corrosion rates (0.21 and 0.38 mm/year, respectively) than the crystalline alloy (0.85 mm/year). The higher corrosion resistance of the BMGs is attributed to their single-phase, chemically homogeneous alloy system, and to the absence of dislocations, grain boundaries or deleterious secondary phases, which may localize dissolution [275,276]. In addition, the reduced corrosion rates can also be ascribed to the anticorrosion-favorable elements, such as zinc, with a high content exceeding its solubility in an equilibrium state in the alloys. The role of Zn in the electrochemical behavior of Mg–Zn–Ca alloys can be explained through its Pourbaix diagram [7]. At pH 7.4 and at lower potentials, Zn is stable in its metallic/elemental form, and thus may not contribute to the formation of any oxide layer. However, at higher potentials, Zn exists in its noninert state, and thus being reactive it detaches and dissolves as Zn^{2+} ions, subsequently forming solid precipitates, such as Zn hydroxide [7], that, together with magnesium oxide/hydroxide, leads to the formation of a double oxide/hydroxide, providing a continuous and uniform coverage of the Mg–Zn–Ca samples' surface. Nowosielski and Cesarz-Andraczke [277] studied the corrosion behavior of different Mg–Zn–Ca BMGs, with Zn content varying from 28 at.% to 32 at.%, and reported that the smallest corrosion rate was obtained by the alloys with 32 at.% Zn. However, the corrosion rate of Mg–Zn–Ca BMGs is still too high for biomedical applications and thus it is necessary to work on the enhancement of their corrosion behavior. In this regard, some studies in literature deal with adding ductile metallic particles, but the results are still debatable. Wong et al. [278] reported that the corrosion resistance of $Mg_{60}Zn_{35}Ca_5$ improved by adding 40 vol.% Ti particles, whereas it was decreased for $Mg_{67}Zn_{28}Ca_5$ by adding the same amount of Ti particles. Other researchers have instead studied the effect of minor additions of alloying elements. However, the addition of each minor element has a specific impact on the degradation behavior since it can lead to the presence of nanocrystals, secondary phases and grain boundaries that create a heterogeneous surface. Qin et al. reported that minor addition of 1 at.% Ag improves the corrosion resistance of $Mg_{65}Zn_{30}Ca_5$ BMGs, whereas the addition of 3 at.% Ag leads to a lower glass-forming ability, inducing the formation of Mg_7Zn_3 second phases that decrease the corrosion resistance. Further data are reported in the following tables (see Table 3.25 for the potentiodynamic polarization curves and Table 3.26 for the corrosion rate data), including the addition of other particles/elements.

3.4.4.3   Mg–Zn–Ca BMGs: Biocompatibility

Since calcium ions are reported to contribute to the control of microtubule assemblies in growth cones and guidance of growth cone extension [279,280] and zinc is reported to play a role as a signaling substance through the body as a modulator of synaptic activity and to inhibit bacterial growth at the surgical site [281,282], Mg–Zn–Ca BMGs are considered highly biocompatible and thus they have attracted the researcher's. Gu et al. showed that MG63 cells adhered well and proliferated on the surface of $Mg_{66}Zn_{30}Ca_4$ [267], in accordance with the study of Li et al. [283] that observed well-adhered and healthy cells of MC3T3-E1 on the surface of $Mg_{66}Zn_{30}Ca_3Sr_1$. In addition, Gu et al. [267] reported the BMGs to be characterized by a better cellular

TABLE 3.25

Summary of the DC polarization curve results for Mg–Zn–Ca BMGs.*

Zn content (at.%)	Ca content (at.%)	Additional elements	Corrosive environment	E_{corr} (V)	i_{corr} ($\mu A/cm^2$)	Ref.
30	4	—	SBF, 37°C	−1.20	3.5	[267]
25	5	—	SBF, 37°C	−1.28	11.2	[267]
28	3	—	Ringer's, 37°C	−1.27	62	[277]
28	4	—	Ringer's, 37°C	−1.35	41	[277]
28	5	—	Ringer's, 37°C	−1.26	55	[277]
28	6	—	Ringer's, 37°C	−1.21	76	[277]
30	3	—	Ringer's, 37°C	−1.26	30	[277]
30	4	—	Ringer's, 37°C	−1.34	29	[277]
30	5	—	Ringer's, 37°C	−1.21	28	[277]
30	6	—	Ringer's, 37°C	−1.20	33	[277]
32	3	—	Ringer's, 37°C	−1.32	21	[277]
32	4	—	Ringer's, 37°C	−1.32	24.6	[277]
32	5	—	Ringer's, 37°C	−1.27	24.7	[277]
32	6	—	Ringer's, 37°C	−1.18	40	[277]
23	4	—	SBF, 37°C	−1.30	—	[268]
34	6	—	SBF, 37°C	−1.17	—	[268]

* All the Mg–Zn–Ca BMGs are cast, unless otherwise specified. Additional information can be found on www.routledge.com/9780367429454.

response than as-rolled pure Mg (Figure 3.11) due to a better corrosion resistance. In fact, the as-rolled pure Mg was characterized by the peeling off of some corrosion layer, which is reported to weaken the adhesion of cells [161], whereas very less peeling off of the corrosion layer is seen for the BMGs because of the combined protection of the magnesium hydroxide and zinc hydroxide.

Further results confirming the cytocompatibility of Mg–Zn–Ca BMGs are reported in Table 3.27.

3.4.5   Mg–Zn–Mn Alloys

Mg–Zn–Mn magnesium alloys (ZM) have the advantages of age hardening, excellent processing performance, good corrosion resistance and low cost due to no noble metal elements (e.g., Zr and rare elements). In fact, Zn as a major alloying element in magnesium alloys has a marked response to precipitation hardening due to the formation of Mg–Zn intermetallic compounds. These compounds are more stable than $Mg_{17}Al_{12}$ at elevated temperature, resulting in higher mechanical properties at ambient and elevated temperatures than Mg–Al alloys [284–286], so Zn was found to be better than aluminum in strengthening effectiveness as an alloying element in magnesium [287]. In

TABLE 3.26

Summary of the corrosion rates for Mg–Zn–Ca BMGs.*

Zn content (at.%)	Ca content (at.%)	Additional elements (at.%)	Corrosive environment	Immersion time	Corrosion rate (mm/year)	Procedure	Ref.
30	4	—	SBF, 37°C	30 days	0.21	WL	[267]
25	5	—	SBF, 37°C	30 days	0.38	WL	[267]
35	5	—	SBF, 37°C	12 weeks	0.61	WL	[278]
28	5	—	SBF, 37°C	11 wees	0.37	WL	[278]
35	5	40 vol. % Ti particles	SBF, 37°C	12 weeks	0.25	WL	[278]
28	5	40 vol. % Ti particles	SBF, 37°C	4 weeks	1.88	WL	[278]
28	3	—	Ringer's, 37°C	1 day	0.56	H	[277]
28	4	—	Ringer's, 37°C	1 day	0.41	H	[277]
28	5	—	Ringer's, 37°C	1 day	0.50	H	[277]
28	6	—	Ringer's, 37°C	1 day	0.64	H	[277]
30	3	—	Ringer's, 37°C	1 day	0.42	H	[277]
30	4	—	Ringer's, 37°C	1 day	0.30	H	[277]
30	5	—	Ringer's, 37°C	1 day	0.44	H	[277]
30	6	—	Ringer's, 37°C	1 day	0.49	H	[277]
32	3	—	Ringer's, 37°C	1 day	0.22	H	[277]
32	4	—	Ringer's, 37°C	1 day	0.19	H	[277]
32	5	—	Ringer's, 37°C	1 day	0.13	H	[277]
32	6	—	Ringer's, 37°C	1 day	0.27	H	[277]

* All the Mg–Zn–Ca BMGs are cast, unless otherwise specified. Additional information can be found on www.routledge.com/9780367429454.

addition, ZM61, subjected to solution and aging treatment, has been reported to achieve mechanical properties close to those of commercial wrought ZK60A alloy [288]. The addition of alloying element Mn to Mg–Zn-based alloy is helpful in refining the grain size and improving its heat resistance and corrosion-resisting properties [289]. Mn has been reported to have the function of refining grain size and improving the tensile strength of Mg alloy [290], to weaken the disadvantageous effects of the impurity Fe on the corrosion properties of magnesium alloys [143] and to be beneficial to the corrosion resistance. Also, the formation of dense film layers of MnO and MnO_2 on the surface can inhibit Cl^- permeation and control the matrix dissolution [291,292]. In addition, *in vivo* study showed that after 18 weeks, about 54% as-cast ZM11 implant had degraded but the degradation of magnesium did not cause any increase in serum magnesium content or did not show any disorders of the kidney after 15-week post implantation [293]. This is because zinc is recognized as a highly essential element for human and manganese has no

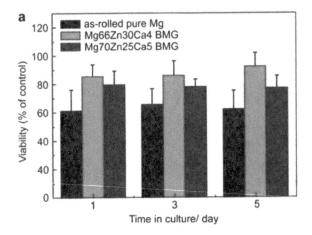

FIGURE 3.11 Cytotoxicity of L929 cells cultured in as-rolled pure Mg, $Mg_{66}Zn_{30}Ca_4$ and $Mg_{70}Zn_{25}Ca_5$ extraction mediums. Reprinted with permission from Elsevier [267].

toxic effect except after extreme occupational exposure and it plays a primary role in the activation of multiple enzyme systems, such as hydrolases, kinases, transferases, decarboxylases and mitochondrial respiration [294].

3.4.5.1 Mg–Zn–Mn Alloys: Mechanical Properties

Although ZM series alloys are attractive for the great biocompatibility of their constituents, the yield strength of the cast Mg–Zn–Mn magnesium alloy is still too low to be applied directly [295]. To increase the mechanical properties, researchers have focused mainly on two different approaches: alloying and plastic deformation technique. Concerning the former, the improvements in strength are due to the second phase strengthening, which is related to the volume fraction of precipitates and the diameter of precipitates; it increases with the increasing volume fraction and the decreasing diameter [296,297]. So, in light of this, the results obtained by Yang et al. [298] are explained. They investigated the effect of Ca addition (in amounts ranging from 0.1 to 1.6 wt.%) on extruded ZM21 alloy, and reported that increasing the Ca amount above 0.33 wt.% leads to a reduction in strength, limited when the Ca amount is 0.7 wt.% and very pronounced when 1.6 wt.% of Ca is added. In the latter case, the reduced mechanical properties are due to the formation of large number of Mg_2Ca phases. They have features of long strip and a long size larger than 15 μm, causing stress concentration and brittle fractures during deformation. Previous studies showed that the strength and ductility decrease rapidly if the size of particle exceeds 10 μm [299,300]. Concerning the improvement of the mechanical properties through plastic deformation techniques, the most investigated was the extrusion process. The extrusion process leads to the formation of a strong basal fiber texture in which the grains are oriented with their basal planes parallel to the extrusion direction that increases the strength of the

TABLE 3.27

Summary of the biocompatibility of Mg–Zn–Ca BMGs.*

Zn content (at. %)	Ca content (at. %)	Test environment	Cell type, procedure	Cell viability		Hemolysis (%)	Ref.
				Time of culture	Result (%)		
30	4	DMEM, 37°C	L929, IC	1 day	85.7	—	[267]
30	4	DMEM, 37°C	L929, IC	3 days	86.2	—	[267]
30	4	DMEM, 37°C	L929, IC	5 days	92.1	—	[267]
25	5	DMEM, 37°C	L929, IC	1 day	79.1	—	[267]
25	5	DMEM, 37°C	L929, IC	3 days	77.8	—	[267]
25	5	DMEM, 37°C	L929, IC	5 days	77.2	—	[267]
30	4	DMEM, 37°C	L929, DC	1 day	92.5	—	[267]
30	4	DMEM, 37°C	L929, DC	3 days	87.1	—	[267]
30	4	DMEM, 37°C	L929, DC	5 days	89.8	—	[267]
25	5	DMEM, 37°C	L929, DC	1 day	85.0	—	[267]
25	5	DMEM, 37°C	L929, DC	3 days	56.6	—	[267]
25	5	DMEM, 37°C	L929, DC	5 days	57.8	—	[267]
30	4	DMEM, 37°C	MG63, IC	1 day	91.1	—	[267]
30	4	DMEM, 37°C	MG63, IC	3 days	90.0	—	[267]
30	4	DMEM, 37°C	MG63, IC	5 days	96.4	—	[267]
25	5	DMEM, 37°C	MG63, IC	1 day	83.9	—	[267]
25	5	DMEM, 37°C	MG63, IC	3 days	83.9	—	[267]
25	5	DMEM, 37°C	MG63, IC	5 days	86.7	—	[267]
30	4	DMEM, 37°C	MG63, DC	1 day	94.2	—	[267]
30	4	DMEM, 37°C	MG63, DC	3 days	87.5	—	[267]
30	4	DMEM, 37°C	MG63, DC	5 days	89.5	—	[267]

25	5	DMEM, 37°C	MG63, DC	1 day	86.6	—	[267]
25	5	DMEM, 37°C	MG63, DC	3 days	83.4	—	[267]
25	5	DMEM, 37°C	MG63, DC	5 days	57.8	—	[267]
35	5	DMEM, 37°C	MG63, IC	1 day	99.6	—	[278]
35	5	DMEM, 37°C	MG63, IC	3 days	106.7	—	[278]
35	5	DMEM, 37°C	MG63, IC	5 days	108.8	—	[278]
28	5	DMEM, 37°C	MG63, IC	1 day	100.6	—	[278]
28	5	DMEM, 37°C	MG63, IC	3 days	106.3	—	[278]
28	5	DMEM, 37°C	MG63, IC	5 days	110.1	—	[278]
35	5 (+ 40 vol. % Ti particles)	DMEM, 37°C	MG63, IC	1 day	90.7	—	[278]
35	5 (+ 40 vol. % Ti particles)	DMEM, 37°C	MG63, IC	3 days	100.2	—	[278]
35	5 (+ 40 vol. % Ti particles)	DMEM, 37°C	MG63, IC	5 days	103.6	—	[278]
28	5 (+ 40 vol. % Ti particles)	DMEM, 37°C	MG63, IC	1 day	105.0	—	[278]
28	5 (+ 40 vol. % Ti particles)	DMEM, 37°C	MG63, IC	3 days	102.9	—	[278]
28	5 (+ 40 vol. % Ti particles)	DMEM, 37°C	MG63, IC	5 days	99.9	—	[278]
30	4	α-MEM, 37°C	MC3T3-E1, IC	1 day	83.0	—	[283]
30	3 (+ 1 at.% Sr)	α-MEM, 37°C	MC3T3-E1, IC	1 day	83.8	—	[283]
30	2.5 (+ 1.5 at.% Sr)	α-MEM, 37°C	MC3T3-E1, IC	1 day	78.7	—	[283]
30	4	α-MEM, 37°C	MC3T3-E1, IC	3 days	105.0	—	[283]
30	3 (+ 1 at.% Sr)	α-MEM, 37°C	MC3T3-E1, IC	3 days	102.1	—	[283]
30	2.5 (+ 1.5 at.% Sr)	α-MEM, 37°C	MC3T3-E1, IC	3 days	105.0	—	[283]
30	4	α-MEM, 37°C	MC3T3-E1, IC	5 days	104.6	—	[283]
30	3 (+ 1 at.% Sr)	α-MEM, 37°C	MC3T3-E1, IC	5 days	103.5	—	[283]
30	2.5 (+ 1.5 at.% Sr)	α-MEM, 37°C	MC3T3-E1, IC	5 days	101.2	—	[283]

* All the Mg–Zn–Ca BMGs are cast, unless otherwise specified. Additional information can be found on www.routledge.com/9780367429454.

alloys. In addition, as all the processes inducing plastic deformations, extrusion leads to the grain refinement due to the dynamic recrystallization. And the dynamic recrystallization is strongly increased by the presence of Mn. In fact, as a result of the low solubility of Mn in Mg matrices (less than 0.1% at a temperature of 300°C [301]), the addition of relatively high amounts of Mn in Mg–Zn-based extruded alloys is likely to generate many fine Mn precipitates (the size of the spherical Mn particle is about 65 nm [302]) that are known to effectively restrict the growth of dynamically recrystallized (DRXed) grains and facilitate the formation of fine DRXed grains during extrusion [301]. In addition, the presence of a large number of fine Mn precipitates effectively hinders the dislocation and leads to its accumulation, increasing the strength.

Concerning the optimum content of Zn, Zhang et al. [303] have studied its effect (4–9%, in wt.%) on microstructures and mechanical properties of the extruded Mg–Zn–Mn magnesium alloys and have found that the as-extruded Mg–5Zn–1Mn (ZM51) alloy achieves the optimal comprehensive mechanical properties, which disagrees with the results found in ref. [304] that reported Mg–Zn–Mn alloy with about 3 wt.% Zn and 1 wt.% Mn to have optimal mechanical properties.

In Table 3.28, the mechanical properties of the investigated Mg–Zn–Mn alloys are reported, reporting also the effect of thermal treatments in improving the strength of the alloys, in particular Mg–6Zn–1Mn (ZM61) alloy exhibits the highest strength of up to 366 MPa but poor ductility after solution and two-step aging treatment (T6) [304].

3.4.5.2 Mg–Zn–Mn Alloys: Corrosion Resistance

Zinc and manganese help to overcome the harmful corrosive effect of impurities such as iron and nickel and those of other heavy-metal elements, and thus improve the corrosion resistance [141]. In addition, the formation of dense passive film layers of $Zn(OH)_2$, ZnO, MnO and MnO_2 on the surface can inhibit the chloride-ion permeation and control matrix dissolution. This can be seen from the electrochemical characterization (polarization curves) of the alloy. Rosalbino et al. [305] studying three different Mg–Zn–Mn alloys reported that the anodic polarization curve of these alloys exhibits a current plateau, which significantly increases with increasing the Zn and Mn content. As a matter of fact, the current plateau is over 180 mV for the Mg–1.5Zn–0.5Mn alloy and more than 270 mV for the Mg–1.5Zn–1Mn alloy (Figure 3.12).

Therefore, increasing Zn and Mn content in the alloy progressively reduces the surface reactivity of magnesium and favors the formation of corrosion products with improved protective capability. However, if the solid solubility limit of the two elements is exceeded, second phases (α-Mn and Mg_7Zn_3) would form, leading to the onset of a galvanic cell with the α-Mg matrix. Although in some studies it is stated that coarse and numerous second phases can act as obstacle to hinder the propagation of corrosion [298], it is widely reported that the lower the second phases volume fractions, the weaker the microgalvanic couple effect [306]. In fact, the lowest corrosion rate in literature has been found for the alloy characterized by the lowest amount of second phases

TABLE 3.28

Summary of the mechanical properties of Mg–Zn–Mn alloys.*

Alloy	Mn content (wt. %)	Additional element (wt %)	Yield strength (MPa)	UTS (MPa)	Elongation to failure (%)	Compressive yield strength (MPa)	Ultimate compressive strength (MPa)	Ref.
ZM51 (homogenized + extruded)	1.06	—	229.9	299.9	11.5	—	—	[289]
ZM21 (homogenized + extruded)	1	0.3 Sr	257	298	21.1	196	386.4	[301]
ZM22 (homogenized + extruded)	2	0.3 Sr	312	333	17	240	425.1	[301]
ZM61 (homogenized + extruded)	1	—	207.3	303.2	11.3	—	—	[302]
ZM61 (homogenized + extruded)	1	0.7 Y	229.6	327.2	9.8	—	—	[302]
ZM61 (homogenized + extruded)	1	2 Y	267.2	328.9	9.3	—	—	[302]
ZM61 (homogenized + extruded)	1	3 Y	277.5	335.8	10.0	—	—	[302]
ZM61 (homogenized + extruded)	1	6 Y	345.2	388.9	6.3	—	—	[302]
ZM11 (extruded)	1		246.5	280.3	21.8	—	—	[304]
ZM21 (extruded)	1		248.8	283.8	20.9	—	—	[304]
ZM31 (extruded)	1		275.9	315.5	10.5	—	—	[304]
ZM21 (homogenized + extruded)	0.94	—	157	251	15.2	—	—	[298]
ZM21 (homogenized + extruded)	1.05	0.1 Ca	188	262	22.2	—	—	[298]
ZM21 (homogenized + extruded)	0.99	0.3 Ca	200	265	19.6	—	—	[298]
ZM21 (homogenized + extruded)	1.07	0.7 Ca	187	260	21.5	—	—	[298]
ZM21 (homogenized + extruded)	0.93	1.6 Ca	177	257	14.6	—	—	[298]
ZM61 (homogenized + extruded)	1	—	209	305	11.5	—	—	[295]
ZM61 (homogenized + extruded)	1	0.5 Ce	232	304	14.7	—	—	[295]
ZM61 (homogenized + extruded + T6)	1	—	314	352	7.9	—	—	[295]
ZM61 (homogenized + extruded + T6)	1	0.5 Ce	267	297	9.3	—	—	[295]

(Continued)

TABLE 3.28 (Cont.)

Alloy	Mn content (wt. %)	Additional element (wt %)	Yield strength (MPa)	UTS (MPa)	Elongation to failure (%)	Compressive yield strength (MPa)	Ultimate compressive strength (MPa)	Ref.
ZM61 (homogenized + extruded + T4 + double aged)	1	—	338	366	5.2	—	—	[295]
ZM61 (homogenized + extruded + T4 + double aged)	1	0.5 Ce	287	308	10.4	—	—	[295]
ZM41 (homogenized + extruded)	0.95	—	239	299	11.6	—	—	[303]
ZM51 (homogenized + extruded)	0.93	—	235	312	13.3	—	—	[303]
ZM61 (homogenized + extruded)	0.94	—	209	305	11.6	—	—	[303]
ZM71 (homogenized + extruded)	0.93	—	222	311	13.1	—	—	[303]
ZM81 (homogenized + extruded)	0.96	—	227	320	13	—	—	[303]
ZM91 (homogenized + extruded)	0.97	—	225	327	14.4	—	—	[303]
ZM41 (homogenized + extruded + T4)	0.95	—	235	278	7	—	—	[303]
ZM51 (homogenized + extruded + T4)	0.93	—	262	304	6.9	—	—	[303]
ZM61 (homogenized + extruded + T4)	0.94	—	314	352	8.0	—	—	[303]
ZM71 (homogenized + extruded + T4)	0.93	—	334	363	3.5	—	—	[303]
ZM81 (homogenized + extruded + T4)	0.96	—	331	349	0.4	—	—	[303]
ZM91 (homogenized + extruded + T4)	0.97	—	343	374	2.6	—	—	[303]
ZM41 (homogenized + extruded + T4 + double aged)	0.95	—	243	284	6.0	—	—	[303]
ZM51 (homogenized + extruded + T4 + double aged)	0.93	—	293	325	4.8	—	—	[303]
ZM61 (homogenized + extruded + T4 + double aged)	0.94	—	338	366	5.2	—	—	[303]

ZM71 (homogenized + extruded + T4 + double aged)	0.93	—	343	370	4.3	—	—	[303]
ZM81 (homogenized + extruded + T4 + double aged)	0.96	—	356	382	1.9	—	—	[303]
ZM91 (homogenized + extruded + T4 + double aged)	0.97	—	355	378	1.7	—	—	[303]
ZM61 (homogenized + extruded)	1.25	—	213	312	11.1	—	—	[288]
ZM61 (homogenized + extruded + T4)	1.25	—	192	297	10.6	—	—	[288]
ZM61 (homogenized + extruded + T6)	1.25	—	281	333	7.8	—	—	[288]
ZM61 (homogenized + extruded + T4 + double aged)	1.25	—	340	366	6.3	—	—	[288]

* All the Mg–Zn–Mn alloys are cast, unless otherwise specified. Additional information can be found on www.routledge.com/9780367429454

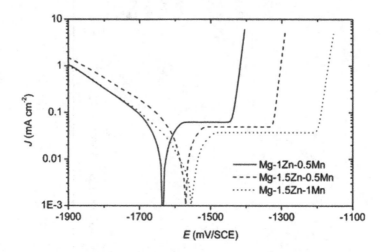

FIGURE 3.12 Polarization curves of Mg–Zn–Mn in Ringer's physiological solution at 37°C. Reprinted with permission from Wiley [305].

[15] (see Tables 3.29 and 3.30 for comparison with corrosion rate from other samples). In addition, an improvement in the corrosion resistance can be obtained by refining the grain size because the large number of grain boundaries can be used as a physical barrier to block the corrosion path [307,308].

3.4.5.3 Mg–Zn–Mn Alloys: Biocompatibility

As previously reported, zinc and manganese are recognized as highly essential elements for human and they play a primary role in the activation of multiple enzyme system [294]. With this in mind, the Grade 0–1 cytotoxicity obtained for homogenized Mg–2Zn–0.2Mn alloys by Zhang et al. [309] through the direct culture method and by others (see Table 3.31) can be easily understood. Moreover, also *in vivo* studies confirm the biocompatibility of Mg–Zn–Mn alloys. In particular, Xu et al. implanted Mg–1.2Mn–1.0Zn implants into rats and observed no increase in serum magnesium content or any disorders of the rats' kidneys after 15-weeks post implantation[293]. Zhang et al. [310], despite reporting a good cytocompatibility of extruded ZM11, found a hemolysis rate of about 65.75%. According to ISO 10993-4 standard, the hemolysis rate of the materials to be used in blood environment has to be less than 5% to avoid any destructive effect of medical materials to the erythrocyte. Zhang et al. ascribed the hemolysis to the increasing pH value, not to the increasing of alloying element (Mg^{2+}, Mn^{2+}, Zn^{2+}) concentration in the extract liquid.

TABLE 3.29

Summary of the polarization curve results for Mg–Zn–Mn alloys.*

Alloy	Mn content (wt.%)	Additional elements	Corrosive environment	E_{corr} (V)	i_{corr} (µA/cm^2)	Ref.
ZM11 (extruded)	1	—	Hank's, 37°C	−1.47	—	[304]
ZM21 (extruded)	1	—	Hank's, 37°C	−1.46	—	[304]
ZM31 (extruded)	1	—	Hank's, 37°C	−1.50	—	[304]
ZM21 (homogenized + extruded)	0.94	—	3.5 wt.% NaCl	−1.59	75.7	[298]
ZM21 (homogenized + extruded)	1.05	0.1 Ca	3.5 wt.% NaCl	−1.63	83.0	[298]
ZM21 (homogenized + extruded)	0.99	0.3 Ca	3.5 wt.% NaCl	−1.62	66.5	[298]
ZM21 (homogenized + extruded)	1.07	0.7 Ca	3.5 wt.% NaCl	−1.59	39.7	[298]
ZM21 (homogenized + extruded)	0.93	1.6 Ca	3.5 wt.% NaCl	−1.64	87.1	[298]
ZM20	0.2	1.1 Ca	SBF, 37°C	−1.77	573.9	[306]
ZM20 (heat treated)	0.2	1.1 Ca	SBF, 37°C	−1.64	260.2	[306]
Mg–1Zn–0.5Mn	0.6	—	Ringer's, 37°C	−1.64	70	[305]
Mg–1.5Zn–0.5Mn	0.6	—	Ringer's, 37°C	−1.57	50	[305]
Mg–1.5Zn–1Mn	1	—	Ringer's, 37°C	−1.55	40	[305]

* All the Mg–Zn–Mn alloys are cast, unless otherwise specified.

3.4.6 MG–ZN–RE ALLOYS

By RE elements we refer to an element from a set of 17 chemical elements in the periodic table, specifically the 15 lanthanides, as well as scandium (Sc) and yttrium (Y). However, the most common magnesium alloying elements are cerium (Ce), lanthanum (La), yttrium (Y), neodimium (Nd), praseodymium (Pr) and gadolinium (Gd). Besides being added as single constituents, RE elements can also be added through Ce-rich MM, which is a compound containing Ce, La, Nd and Pr in different amounts. In literature, three different kinds of ternary equilibrium phases are reported in the Mg–Zn–Y system alloys (ZE alloys, ZW if the RE element is yttrium): I-phase, W-phase and Z- or X-phase, and their appearance and volume fraction can be controlled by adjusting the mass/molar ratio of Zn/Y [311]. I-phase (Mg_3Zn_6Y) is a quasicrystalline icosahedral phase [312] that forms when the Zn/Y weight ratio is 5–7 according to ref. [311] or higher than 4.38 according to ref. [313]. This icosahedral phase possesses five-fold symmetry and a quasiperiodic structure, which is very different from crystalline phases. In addition, this kind of phase is coherent with the Mg matrix [314], showing a definite orientation relationship and with a strong interface [315]. W-phase ($Mg_3Zn_3Y_2$) is a dendritic phase with a face center cubic

TABLE 3.30

Summary of the corrosion rates for Mg–Zn–Mn alloys.*

Alloy	Mn content (wt.%)	Additional elements (wt.%)	Corrosive environment	Immersion time	Corrosion rate (mm/ year)	Procedure	Ref.
ZM20	0.2	—	Hank's, 37°C	17 days	0.026	H	[15]
ZM11 (extruded)	1	—	Hank's, 37°C	9 days	0.039	WL	[304]
ZM21 (extruded)	1	—	Hank's, 37°C	9 days	0.11	WL	[304]
ZM31 (extruded)	1	—	Hank's, 37°C	9 days	0.44	WL	[304]
ZM21 (homogenized + extruded)	0.94	—	3.5 wt.% NaCl	3 days	1.99	WL	[298]
ZM21 (homogenized + extruded)	1.05	0.1 wt.% Ca	3.5 wt.% NaCl	3 days	1.86	WL	[298]
ZM21 (homogenized + extruded)	0.99	0.3 wt.% Ca	3.5 wt.% NaCl	3 days	1.75	WL	[298]
ZM21 (homogenized + extruded)	1.07	0.7 wt.% Ca	3.5 wt.% NaCl	3 days	1.54	WL	[298]
ZM21 (homogenized + extruded)	0.93	1.6 wt.% Ca	3.5 wt.% NaCl	3 days	3.0	WL	[298]

* All the Mg–Zn–Mn alloys are cast, unless otherwise specified. Additional information can be found on www.routledge.com/9780367429454.

(fcc) structure with a lattice parameter of 0.685 nm [316]. Although the elastic modulus and hardness of W-phase are higher than that of α-Mg [317,318], it is a hard and brittle phase, which is generally not conducive to the deformation and to the strengthening effect in Mg–Zn–Y alloys since most of the existing investigations reported that it was not coherent with Mg matrix and the interface was easy to debond [319–322]. For the Zn/Y ratio inducing to the formation of W-phase, Xu et al. [313] stated that when the Zn/Y ratio is between 1.10 and 4.38, both I-phase and W-phase could be simultaneously formed, whereas Lee et al. [311] stated that α-Mg + I-phase +W-phase are present when the Zn/Y weight ratio is 2.5–2, and α-Mg+W-phase when the Zn/Y weight ratio is 2–1.5. Concerning the Z- or X-phase ($Mg_{12}ZnY$), it is a long-period stacking ordered (LPSO) phase with periodic stacking of its (0001) basal plane along the c-axis [323–325]. LPSO structures are coherent with α-Mg matrix being an isotope of the hexagonal close-packed crystallographic structure of Mg crystals

TABLE 3.31
Summary of the biocompatibility of Mg–Zn–Mn alloys.*

| Alloy | Additional elements | Test environment | Cell viability | | | Hemolysis (%) | Ref. |
			Cell type, procedure	Time of culture	Result (%)		
ZM20 (homogenized)	—	DMEM, 37°C	BMSCs, DC	1 day	87.5	—	[309]
ZM20 (homogenized)	0.38 wt.% Ca	DMEM, 37°C	BMSCs, DC	1 day	91.1	—	[309]
ZM20 (homogenized)	0.76 wt.% Ca	DMEM, 37°C	BMSCs, DC	1 day	90.3	—	[309]
ZM20 (homogenized)	1.1 wt.% Ca	DMEM, 37°C	BMSCs, DC	1 day	90.3	—	[309]
ZM20 (homogenized)	—	DMEM, 37°C	BMSCs, DC	2 days	96.8	—	[309]
ZM20 (homogenized)	0.38 wt.% Ca	DMEM, 37°C	BMSCs, DC	2 days	101.2	—	[309]
ZM20 (homogenized)	0.76 wt.% Ca	DMEM, 37°C	BMSCs, DC	2 days	100.7	—	[309]
ZM20 (homogenized)	1.1 wt.% Ca	DMEM, 37°C	BMSCs, DC	2 days	100.9	—	[309]
ZM20 (homogenized)	—	DMEM, 37°C	BMSCs, DC	3 days	97.4	—	[309]
ZM20 (homogenized)	0.38 wt.% Ca	DMEM, 37°C	BMSCs, DC	3 days	109.0	—	[309]
ZM20 (homogenized)	0.76 wt.% Ca	DMEM, 37°C	BMSCs, DC	3 days	102.1	—	[309]
ZM20 (homogenized)	1.1 wt.% Ca	DMEM, 37°C	BMSCs, DC	3 days	101.8	—	[309]
ZM11 (extruded)	—	DMEM, 37°C	L-929, IC	1 day	100.0	—	[310]
ZM11 (extruded)	—	DMEM, 37°C	L-929, IC	2 days	124.1	—	[310]
ZM11 (extruded)	—	DMEM, 37°C	L-929, IC	3 days	100.0	—	[310]
ZM11 (extruded)	—	PBS with diluted blood, 37°C	—	—	—	65.8	[310]

* All the Mg–Zn–Mn alloys are cast, unless otherwise specified. Additional information can be found on www.routledge.com/9780367429454.

and exhibit rhombohedral (R) and hexagonal (H) Bravais lattices, depending on the stacking period of the close-packed atomic layers [326]. Based on stacking sequence, typical LPSO structures include 6H, 10H, 12H, 14H, 15R, 18R, 21R and 24R [326,327]. Among them, the 18R structure could be formed in either as-cast or deformed conditions, which could be transformed into 14H structure during heat treatment at 350–500°C [328,329]. It is noteworthy that Mg–Zn–Y alloys were generally designed with a constant Zn/Y ratio of 0.5 to form LPSO phase [330–332]. LPSO phase can be formed in other Mg–Zn–RE alloys, if the criteria reported in ref. [333] are met. However, although Mg–Zn–RE alloys have been widely studied, their applications in the biomedical field are debatable since some RE elements may induce latent toxic and harmful effects on the human body during degradation [151,334]. Highly concentrated RE elements were observed in different inner organs in rabbits after long-term implantation of RE-containing magnesium alloys [126]. For example, high amounts of Y (above its daily intake of 4.2 mg/day [335]) can cause increases of blood eosinocyte, decreases of body weight and eosinophil infiltration in the submucosa [336]. However, another study indicated that Mg alloys containing small amounts of yttrium and RE elements would be appropriate for biomedical applications [337]. So it is necessary to further investigate the effect of the addition of Y in Mg alloys on biocompatibility.

3.4.6.1 Mg–Zn–RE Alloys: Mechanical Properties

The I-, W- and LPSO phases have different effects on the mechanical properties and most of the works deal with their influence on strength and elongation. I- and LPSO phases are closely bonded with the Mg matrix and can effectively retard the basal slip, thus strengthening the alloy greatly. Concerning the I-phase, the high strength of rolled Mg–3Zn–0.6Y and rolled Mg–6Zn–1.2Y reported in ref. [338] is mainly attributed to the presence of a large number of fine and uniformly distributed I-phase particles in the -Mg matrix, which can impede the dislocation movement and propagation. Moreover, grain size of Mg–Zn–Y alloys decreases with increasing the volume fraction of I-phase particles. Generally, the elongation becomes low in alloys containing a large number of intermetallic particles since dislocations are required to form in the region surrounding the hard particles due to the geometrical effect during deformation, eventually leading to decohesion of the particles from the matrix. However, Mg–6Zn–1.2Y alloy exhibits larger elongation despite having a larger fraction of I-phase particles. It has been reported that the interface between I-phase and α-Mg matrix is so stable that the debonding and micro-scale defect cannot be generated [311]. In addition, the large elongation is also attributed to the weak texture. I-phase can in fact take various orientation relationships with α-Mg grains in Mg–Zn–Y alloy [339,340], which may play a role in randomizing the texture. The relative random texture in Mg–Zn–Y alloys is attributed to local lattice rotations around I-phase particles during shear deformation. Compared to the LPSO structures, the I-phase is cheaper because of the lower content in alloying RE elements, but it is less strengthening: UTS around 450 MPa for extruded $Mg_{96.27}Zn_{3.3}Y_{0.43}$ containing the I-phase [341] and close to 600 MPa for RS/PM $Mg_{97}Zn_1Y_2$ reinforced by the LPSO phase [342]. However, I-phase is characterized by higher ductility, that is, about 15% for $Mg_{96.27}Zn_{3.3}Y_{0.43}$ instead of 5% for $Mg_{97}Zn_1Y_2$. The strengthening effect of LPSO phase is due to its strong basal texture. In fact, the TEM characterization performed in ref. [343] revealed that the LPSO phase elongated along the direction of extrusion had a strong basal texture in which the *c*-axes of the grains were perpendicular to the direction of extrusion (Figure 3.13).

This basal fiber texture is very suitable for the strengthening of the alloy since the plate-like interface of LPSO phases is parallel to the (001) basal plane, which is the dominant deformation mode of Mg alloys. As a result, the LPSO phases inhibit basal slips and the activation of nonbasal slips needs much higher stress. In addition, twinning was not found in the LPSO phase-containing Mg–Zn–Y extruded alloys, while some kink-deformation bands were observed, further strengthening the Mg matrix [344]. However, the strengthening effect of the LPSO structures depends also on their shape, and Cheng et al. [345] reported the following strengthening efficiency trend: block-like LPSO > block-like LPSO + lamellar-like LPSO > rod-shaped LPSO > rod-shaped LPSO + lamellar-like LPSO. Concerning the W-phase, its influence is still debated. Although the majority of the works reports that W-phase containing alloys have relatively low mechanical properties [313,319–321] since the limited symmetry of this phase leads to a noncoherent W-phase/α-matrix interface that can be easily cracked. Feng et al. [346] reported that the tensile strength of the as-cast Mg–Zn–Y–Nd alloy decreases continually with the increasing W-phase. However, some investigations indicate that the W-phase is

FIGURE 3.13 TEM observation of the LPSO structure in a DRXed α-Mg grain. Reprinted with permission from Elsevier [343].

beneficial for improving mechanical properties. For example, Chen et al. [347] reported that the W-phase is not a harmful phase, leading to an increase in strength and elongation. Xu et al. [348] reported that when the volume fraction of W-phase is between 11.2% and 17.5%, the alloys have superior strength due to the strong bonding interface between W-phase and Mg matrix.

Furthermore, from Table 3.32 it can be seen how wrought or severely plastic deformed Mg–Zn–RE alloys are generally characterized by great ductility, due to a more randomized texture achieved after the recrystallization process, and different mechanisms have been proposed in literature. Mishra et al. [349] and Ball and Prangnell [350] suggested that the weaker textures in RE-containing Mg alloys are caused by particle-stimulated nucleation and boundary pinning at Mg–RE precipitates. Mackenzie and Pekguleryuz [351] deduced that a change in the mobility of high-angle boundaries by alloying additions, in the form of a solute drag effect, leads to texture weakening as well as the development of prominent rotated basal component. Nucleation at shear bands formed during the deformation process and a consequent weak texture has been discussed by Stanford and Barnett [352], reporting that RE texture component with <11–21> oriented towards the extrusion direction was a typical characteristic for Mg alloys containing RE. In particular, Gd, Y, La, Ce and Nd are reported to have the highest effect on the texture [353].

3.4.6.2 Mg–Zn–RE Alloys: Corrosion Resistance

Pérez et al. [354] investigated the effect of the addition of Ce MM on Mg–Zn–Y alloys, and they reported that increasing the amount of Ce MM leads to an increase of LPSO second phases, which determines a reduction of the corrosion resistance. The same harmful effect of the LPSO phase on the corrosion resistance has also been reported by Zhao et al. [355]. They studied the corrosion behavior of three different types of Mg–Zn–Y alloys, namely $Mg_{97}Zn_1Y_2$, $Mg_{94}Zn_2Y_4$ and $Mg_{97}Zn_3Y_6$, characterized by different amounts of LPSO phases;25%, 48% and 66%, respectively. Their results report that the onset time of pitting decreases from 5 days of immersion for the $Mg_{97}Zn_1Y_2$ alloy to 1 day of immersion for the $Mg_{97}Zn_3Y_6$ alloy, revealing that increasing the volume fraction of LPSO phase made localized corrosion more susceptive. However, Li et al. reported a beneficial effect of the LPSO phase on the corrosion resistance [356]. I-phase, instead, is reported to improve the corrosion properties because of its low interfacial energy [357]. In addition, Zhang et al. [358] reported that the corrosion property of the alloys with I-phase or W-phase separately is superior to that of the alloys with both I-phase and W-phase. However, the effect of the second phases on the corrosion strongly depends on their network distribution. In fact, ref. [359] reports the LPSO phase with a morphology of a continuous network in the grain boundary region can act as a corrosion barrier to protect the matrix, whereas when the LPSO phase is composed of isolated islands and a lamellar region, the hindering effect is lost. This is due to the passivation stage shown by Y-containing alloys [358]. A protective Y_2O_3 layer can be formed on the alloy surface during the corrosion process [360,361]. In addition, fine and dispersed second

TABLE 3.32

Summary of the mechanical properties of Mg–Zn–RE alloys.*

Alloy	RE and additional elements	Yield strength (MPa)	UTS (MPa)	Elongation to failure (%)	Compressive yield strength (MPa)	Ultimate compressive strength (MPa)	Ref.
Mg–3Zn–0.6Y (rolled + annealed)	0.6 wt.% Y	136	236	26.3	—	—	[338]
Mg–6Zn–1.2Y (rolled + annealed)	1.2 wt.% Y	137	258	30.7	—	—	[338]
Mg₉₇Zn₁Y₂ (extruded)	2 at.% Y	348.0	382.7	7.5	—	—	[343]
Mg–2Zn–0.6Y–0.6 Nd	0.6 wt.% Y + 0.6 wt.% Nd	116	200	19.8	—	—	[346]
Mg–3Zn–0.9Y–0.9 Nd	0.9 wt.% Y + 0.9 wt.% Nd	121	189	13.9	—	—	[346]
Mg–4Zn–1.2Y–1.2 Nd	1.2 wt.% Y + 1.2 wt.% Nd	119	180	12.5	—	—	[346]
Mg–2Zn–0.6Y–0.6 Nd (aged)	0.6 wt.% Y + 0.6 wt.% Nd	134	236	12.1	—	—	[346]
Mg–3Zn–0.9Y–0.9 Nd (aged)	0.9 wt.% Y + 0.9 wt.% Nd	145	248	14.8	—	—	[346]
Mg–4Zn–1.2Y–1.2 Nd (aged)	1.2 wt.% Y + 1.2 wt.% Nd	152	257	13.9	—	—	[346]
Mg–4.86Zn–8.78Y–0.6Ti	8.78 wt.% Y	153	242	6.7	—	—	[345]
Mg–4.86Zn–8.78Y–0.6Ti (heat treated)	8.78 wt.% Y + 0.6 wt.% Ti	147	245	9.5	—	—	[345]
ZW01	1 wt.% Y + 0.5 wt.% Nd + 0.5 wt.% Sn + 0.3 wt.% Zr	—	142.8	9.8	—	—	[347]
ZW11	1 wt.% Y + 0.5 wt.% Nd + 0.5 wt.% Sn + 0.3 wt.% Zr	—	174.7	11.6	—	—	[347]
ZW21	1 wt.% Y + 0.5 wt.% Nd + 0.5 wt.% Sn + 0.3 wt.% Zr	—	207.2	16.9	—	—	[347]
ZW31	1 wt.% Y + 0.5 wt.% Nd + 0.5 wt.% Sn + 0.3 wt.% Zr	—	192.2	13.3	—	—	[347]
ZW41	1 wt.% Y + 0.5 wt.% Nd + 0.5 wt.% Sn + 0.3 wt.% Zr	—	183.4	11.7	—	—	[347]

(Continued)

TABLE 3.32 (Cont.)

Alloy	RE and additional elements	Yield strength (MPa)	UTS (MPa)	Elongation to failure (%)	Compressive yield strength (MPa)	Ultimate compressive strength (MPa)	Ref.
Mg–4Zn–1.6Y (homogenized + extruded)	1 wt.% Y + 0.5 wt.% Nd + 0.5 wt.% Sn + 0.3 wt.% Zr	—	249.5	21.8	—	—	[311]
Mg–4Zn–0.8Y (homogenized + extruded)	1.6 wt.% Y	—	244.3	27.7	—	—	[311]
Mg–4Zn–0.6Y (homogenized + extruded)	0.8 wt.% Y	—	236.5	29.6	—	—	[311]
Mg–3Zn–0.6Y (homogenized + extruded)	0.6 wt.% Y	—	223.6	30.5	—	—	[311]
Mg–6Zn–1.2Y (homogenized + extruded)	0.6 wt.% Y	—	258.1	29.5	—	—	[311]
Mg–8Zn–1.6Y (homogenized + extruded)	1.2 wt.% Y	—	269.6	27.0	—	—	[311]
ZE20 (rolled)	1.6 wt.% Y	—	246.9	6.0	—	—	[353]
ZE20 (rolled + annealed)	0.5 wt.% Ce	—	237.1	13.8	—	—	[353]
ZE20 (pack-rolled)	0.5 wt.% Ce	—	241.0	33.4	—	—	[353]
Mg$_{97}$Zn$_1$Y$_2$ (extruded)	0.5 wt.% Ce	365.3	392.0	7.0	—	—	[333]
Mg$_{97}$Zn$_1$Y$_1$La$_1$ (extruded)	2 at.% Y	383.1	423.2	2.7	—	—	[333]
Mg$_{97}$Zn$_1$Y$_1$Ce$_1$ (extruded)	1 at.% Y+ 1at.% La	380.8	426.5	2.6	—	—	[333]
Mg$_{97}$Zn$_1$Y$_1$Nd$_1$ (extruded)	1 at.% Y+ 1at.% Ce	328.5	356.3	6.9	—	—	[333]
Mg$_{97}$Zn$_1$Y$_1$Sm$_1$ (extruded)	1 at.% Y+ 1at.% Nd	324.1	359.7	9.3	—	—	[333]
Mg$_{90.5}$Zn$_{3.25}$Y$_{6.25}$ (rolled)	1 at.% Y+ 1at.% Sm	353	400	5	—	—	[330]
	6.25 at.% Y						

* All the Mg–Zn–RE alloys are cast, unless otherwise specified. Additional information can be found on www.routledge.com/9780367429454

phases are reported to improve the corrosion resistance and this can be obtained through SPD techniques, such as cyclic extrusion compression, as reported by Wu et al. [362] and gathered in Tables 3.33 and 3.34.

3.4.6.3   Mg–Zn–RE Alloys: Biocompatibility

The majority of works available in literature reported RE elements (e.g., Y, Ce, Nd, Gd, etc.) not to be biocompatible and even toxic to living cells [151,334,363]. Rim et al. [364] reviewed the toxicological effect of RE elements (see Table 3.35).

Few works deal with the assessment of the cytocompatibility of Mg–Zn–RE alloy, and their results are reported in Table 3.36. The majority of these works focuses on the effects of coating procedures in improving the biocompatibility of these alloys [365,366]. The effects of coatings are not the main scope of this chapter (they will be treated in detail in a future publication that will deal with the different coating procedures and their effects on bio-mechanical properties of Mg alloys). Only the results for the bare materials are gathered in Table 3.36. The use of coatings has been investigated to slow down the corrosion rate thereby reducing the cytotoxicity. While Liu et al. [367,368] reported cast and extruded Mg–2Zn–0.5Y–0.5Nd alloy to be cytotoxic for HUVEC, Wang et al. [369] stated the same alloy obtained by cyclic extrusion compression has a good biocompatibility (it is characterized by a Grade 0–1 cytotoxicity). Cyclic extrusion compression leads to an increased corrosion resistance (see Table 3.34). The aim of decreasing the corrosion to improve the biocompatibility is justified by the results obtained in some works that report the RE elements to be toxic only above certain concentrations [335,363]. Song et al. [370] studied the long-term cytotoxicity of Mg–Zn–Y–Nd–Zr alloy by preparing the extracts with a prolonged immersion period of 1440 h, reporting the alloy to be biocompatible when the extract concentrations are lower than the range 3–5% and 5–7% for L929 and MC3T3-E1 cells, respectively.

3.5   MG–CA ALLOYS

Mg–Ca alloys have attracted great attention in the biomedical field. Besides their low density (1.55 g/cm^3), calcium has been shown to be able to improve the mechanical properties of pure Mg (but less than Al or RE elements) due to the formation of thermally stable second phase Mg_2Ca that leads to grain refinement [371,372] and precipitation strengthening [373,374]. Calcium has also been shown to improve corrosion resistance when included in low quantities [375] and, above all, is an essential element in the human body, involved in innumerable reactions including an important role in nerve conduction, muscle contraction, hormone release and cell signaling, and is a primary component of bone (body contains about 1100 g of Ca, nearly 1.5% of the body weight, and 99% of this Ca is contained in the skeleton) [376]. Moreover, after the reduced hydrogen evolution rate obtained either *in vitro* and *in vivo* for K-MET™ Bioresorbable Bone screw (manufactured by U&I and made

TABLE 3.33

Summary of the polarization curve results for Mg–Zn–RE alloys.*

Alloy	RE elements	Additional elements	Corrosive environment	E_{corr} (V)	i_{corr} (μA/ cm^2)	Ref.
Mg$_{95}$Zn$_2$Y$_{1.5}$MM$_{1.5}$ (PM)	1.5 at.% Y + 1.5 at.% MM	—	PBS	−1.39	75	[354]
Mg$_{95}$Zn$_2$Y$_{1.5}$MM$_{1.5}$	1.5 at.% Y + 1.5 at.% MM	—	PBS	−1.46	45	[354]
Mg$_{95}$Zn$_3$Y$_1$MM$_1$	1.5 at.% Y + 1.5 at.% MM	—	PBS	−1.38	15	[354]
ZE41	—	—	Hank's, 37°C	−1.40	9	[21]
Mg$_{97}$Zn$_1$Y$_2$	2 at.% Y	—	SBF, 37°C	−1.54	119	[359]
Mg$_{97}$Zn$_1$Y$_2$	2 at.% Y	—	DMEM + 10% FBS, 37°C	−1.49	7.6	[355]
Mg$_{94}$Zn$_2$Y$_4$	4 at.% Y	—	DMEM + 10% FBS, 37°C	−1.44	2.9	[355]
Mg$_{97}$Zn$_3$Y$_6$	6 at.% Y	—	DMEM + 10% FBS, 37°C	−1.47	3.3	[355]
Mg$_{96.83}$Zn$_1$Y$_2$Zr$_{0.17}$	2 at.% Y	0.17 at.% Zr	DMEM + 10% FBS, 37°C	−1.64	4.5	[355]
Mg$_{96.83}$Zn$_1$Y$_2$Zr$_{0.17}$ (extruded)	2 at.% Y	0.17 at.% Zr	DMEM + 10% FBS, 37°C	−1.70	6.7	[355]
Mg–Zn–Y–Nd	—	—	SBF, 37°C	−1.76	53.0	[357]
Mg–Zn–Y–Nd (rapid solidified)	—	—	SBF, 37°C	−1.57	26.2	[357]
Mg–1.98Zn–0.36Y (extruded)	0.36 wt.% Y	—	Hank's, 37°C	−1.50	1.88	[358]
Mg–1.84Zn–0.82Y (extruded)	0.82 wt.% Y	—	Hank's, 37°C	−1.57	4.47	[358]
Mg–1.73Zn–1.54Y (extruded)	1.54 wt.% Y	—	Hank's, 37°C	−1.53	2.83	[358]

* All the Mg–Zn–RE alloys are cast, unless otherwise specified.

of Mg–Ca alloy) [232], in 2015 it was approved for clinical protocol from the Ministry of Food and Drug Safety in Korea [377].

3.5.1 MG–CA ALLOYS: MECHANICAL PROPERTIES

As mentioned earlier, Ca addition tends to improve the strength of pure Mg due to the decrease in the grain size, solution strengthening effects and second phase strengthening. Li et al. [378] investigated the compressive behavior of as-cast Mg–xCa (x=0.5, 1, 2, 5, 10, 15, 20 wt.%), and they reported that the compressive strength increases with Ca content. In addition, Jeong and Kim

TABLE 3.34

Summary of the corrosion rates for Mg–Zn–RE alloys.*

Alloy	RE elements	Additional elements	Corrosive environment	Immersion time	Corrosion rate (mm/year)	Procedure	Ref.
ZE41	—		Hank's, 37°C	5 days	0.09	H	[21]
ZE41	—		Hank's, 37°C	10 days	0.53	H	[21]
ZE41	—		Hank's, 37°C	12 days	0.67	H	[21]
ZE41	—		Hank's, 37°C	12 days	2.1	WL	[21]
$Mg_{97}Zn_1Y_2$	2 at.% Y	—	SBF, 37°C	3 days	2.1	WL	[359]
$Mg_{97}Zn_1Y_2$	2 at.% Y		SBF, 37°C	5 days	2.9	WL	[359]
$Mg_{94}Zn_1Y_2$	2 at.% Y		DMEM + 10% FBS, 37°C	21 days	2.2	WL	[355]
$Mg_{94}Zn_2Y_4$	4 at.% Y		DMEM + 10% FBS, 37°C	7 days	16.8	WL	[355]
$Mg_{97}Zn_3Y_6$	6 at.% Y	—	DMEM + 10% FBS, 37°C	7 days	32.5	WL	[355]
$Mg_{96.83}Zn_1Y_2Zr_{0.17}$	2 at.% Y	0.17 at.% Zr	DMEM + 10% FBS, 37°C	21 days	1.1	WL	[355]
$Mg_{96.83}Zn_1Y_2Zr_{0.17}$ (extruded)	2 at.% Y	0.17 at.% Zr	DMEM + 10% FBS, 37°C	21 days	0.9	WL	[355]
Mg–Zn–Y–Nd	—		SBF, 37°C	1 day	2.3	WL	[357]
Mg–Zn–Y–Nd (rapid solidified)	—		SBF, 37°C	1 day	1.87	WL	[357]
Mg–Zn–Y–Nd	—		SBF, 37°C	3 days	5.1	WL	[357]
Mg–Zn–Y–Nd (rapid solidified)	—		SBF, 37°C	3 days	4.21	WL	[357]
Mg–Zn–Y–Nd	—		SBF, 37°C	5 days	4.18	WL	[357]
Mg–Zn–Y–Nd (rapid solidified)	—		SBF, 37°C	5 days	3.72	WL	[357]
Mg–Zn–Y–Nd	—		SBF, 37°C	7 days	3.24	WL	[357]
Mg–Zn–Y–Nd (rapid solidified)	—		SBF, 37°C	7 days	2.78	WL	[357]
Mg–1.98Zn–0.36Y (extruded)	0.36 wt.% Y	—	Hank's, 37°C	9 days	0.43	WL	[358]
Mg–1.84Zn–0.82Y (extruded)	0.82 wt.% Y	—	Hank's, 37°C	9 days	1.64	WL	[358]
Mg–1.73Zn–1.54Y (extruded)	1.54 wt.% Y	—	Hank's, 37°C	9 days	0.56	WL	[358]

* All the Mg–Zn–RE alloys are cast, unless otherwise specified. Additional information can be found on www.routledge.com/9780367429454.

TABLE 3.35

Summary of toxicological information of RE elements' bio-assessment.*

Symbol	Name	Toxicological information
Sc	Scandium	Elemental scandium is considered nontoxic, and little animal testing of scandium compounds has been done. The half lethal dose levels for scandium (III) chloride for rats have been determined as 4 mg/kg for intraperitoneal and 755 mg/kg for oral administration
Y	Yttrium	Water-soluble compounds of yttrium are considered mildly toxic, while its insoluble compounds are nontoxic. In experiments on animals, yttrium and its compounds cause lung and liver damage. In rats, inhalation of yttrium citrate caused pulmonary edema and dyspnea, while inhalation of yttrium chloride caused liver edema, pleural effusions and pulmonary hyperemia. Exposure to yttrium compounds in humans may cause lung disease
La	Lanthanum	In animals, the injection of lanthanum solutions produces hyperglycemia, low blood pressure, degeneration of the spleen and hepatic alterations. Lanthanum oxide lethal dose in rat oral (>8500 mg/kg), mouse intraperitoneal (530 mg/kg)
Ce	Cerium	Cerium is a strong reducing agent; it ignites spontaneously in air at 65°C to 80°C. Fumes from cerium fires are toxic. Animals injected with large doses of cerium have died due to cardiovascular collapse. Cerium (IV) oxide is a powerful oxidizing agent at high temperatures, and will react with combustible organic materials. Ceric oxide lethal dose in rat oral (5000 mg/kg), dermal (1000–2000 mg/kg), inhalation dust (5.05 mg/l)
Pr	Praseodymium	Praseodymium is of low to moderate toxicity
Nd	Neodymium	Neodymium compounds are of low to moderate toxicity; however, its toxicity has not been thoroughly investigated. Neodymium dust and salts are very irritating to the eyes and mucous membranes, and moderately irritating to the skin. Neodymium oxide lethal dose in rat oral (> 5000 mg/kg), mouse intraperitoneal (86 mg/kg), and neodymium oxide was investigated as mutagen
Pm	Promethium	It is not known what human organs are affected by interaction with promethium; a possible candidate is the bone tissue. No dangers, aside from the radioactivity, have been shown
Sm	Samarium	The total amount of samarium in adults is about 50 μm, mostly in liver and kidneys, and with about 8 μg/l being dissolved in the blood. Insoluble salts of samarium are non-toxic, and the soluble ones are only slightly toxic. When ingested, only about 0.05% of samarium salt is absorbed into the bloodstream, and the remainder is excreted. From the blood, about 45% goes to the liver, and 45% is deposited on the surface of the bones, where it remains for about 10 years; the balance of 10% is excreted
Eu	Europium	There are no clear indications that europium is particularly toxic compared to other heavy metals. Europium chloride nitrate and oxide have been tested for toxicity: europium chloride shows an acute intraperitoneal lethal dose toxicity of 550 mg/kg, and the acute oral lethal dose toxicity is

(*Continued*)

TABLE 3.35 (Cont.)

Symbol	Name	Toxicological information
		5000 mg/kg. Europium nitrate shows a slightly higher intraperitoneal lethal dose toxicity of 320 mg/kg, while the oral toxicity is above 5000 mg/kg
Gd	Gadolinium	As a free ion, gadolinium is highly toxic, but magnetic resonance imaging contrast agents are chelated compounds, and are considered safe enough to be used in most persons. The toxicity depends on the strength of the chelating agent. Anaphylactoid reactions are rare, occurring in approximately 0.03–0.1%
Tb	Terbium	As with the other lanthanides, terbium compounds are of low to moderate toxicity, although their toxicity has not been investigated in detail
Dy	Dysprosium	Soluble dysprosium salts, such as dysprosium chloride and dysprosium nitrate, are mildly toxic when ingested. The insoluble salts, however, are nontoxic. Based on the toxicity of dysprosium chloride to mice, it is estimated that the ingestion of 500 g or more could be fatal to a human
Ho	Holmium	The element, as with other RE, appears to have a low degree of acute toxicity
Er	Erbium	Erbium compounds are of low to moderate toxicity, although their toxicity has not been investigated in detail
Tm	Thulium	Soluble thulium salts are regarded as slightly toxic if taken in large amounts, but the insoluble salts are nontoxic. Thulium is not taken up by plant roots to any extent, and thus does not get into the human food chain
Yb	Ytterbium	All compounds of ytterbium should be treated as highly toxic, because it is known to cause irritation to the skin and eye, and some might be teratogenic
Lu	Lutetium	Lutetium is regarded as having a low degree of toxicity: for example, lutetium fluoride inhalation is dangerous and the compound irritates skin. Lutetium oxide powder is toxic as well if inhaled or ingested. Soluble lutetium salts are mildly toxic, but insoluble ones are not

* Modified from ref. [364].

[2] found that the yield strength of Mg increased from 27.8 MPa to 42.3 and 81.3 MPa with the addition of 0.4 and 1 wt.% of Ca, respectively. These results were in agreement with those obtained by Bakhsheshi-Rad [235], which showed that an addition of 2 wt.% Ca led to an increase of the UTS from 97.5 to 115.2 MPa. However further increasing the Ca content to 4 wt.% drastically reduced the strength to 77.4 MPa. This agrees with the results reported in ref. [233], where the tensile properties of as-cast Mg–xCa with x=1, 2, 3 wt.% were investigated and both the yield and UTS were found to decrease with increasing the amount of Ca. This phenomenon can be readily interpreted by the amount of precipitates (Mg_2Ca) within the Mg–Ca alloys. The brittle Mg_2Ca phase with network structure along the grain boundary acts as crack source, which reduces

TABLE 3.36

Summary of the biocompatibility of Mg–Zn– RE alloys.*

Alloy	RE and additional elements	Test environment	Cell type, procedure	Time of culture	Result (%)	Hemolysis (%)	Ref.
				Cell viability			
Mg–2Zn–0.5Y–0.5Nd (CEC)	0.5 wt.% Y + 0.5 wt.% Nd	PBS + Diluted blood, 37°C	—	—	—	7.3	[369]
Mg–2Zn–0.5Y–0.5Nd (CEC)	0.5 wt.% Y + 0.5 wt.% Nd	DMEM, 37°C	HUVEC, IC	1 day	87.3	—	[369]
Mg–2Zn–0.5Y–0.5Nd (CEC)	0.5 wt.% Y + 0.5 wt.% Nd	DMEM, 37°C	HUVEC, IC	3 days	76.5	—	[369]
Mg–2Zn–0.5Y–0.5Nd (CEC)	0.5 wt.% Y + 0.5 wt.% Nd	DMEM, 37°C	HUVEC, IC	5 days	75.5	—	[369]
Mg–2Zn–0.5Y–0.5Nd (CEC)	0.5 wt.% Y + 0.5 wt.% Nd	DMEM, 37°C	VSMC, IC	1 day	89.2	—	[369]
Mg–2Zn–0.5Y–0.5Nd (CEC)	0.5 wt.% Y + 0.5 wt.% Nd	DMEM, 37°C	VSMC, IC	3 days	85.6	—	[369]
Mg–2Zn–0.5Y–0.5Nd (CEC)	0.5 wt.% Y + 0.5 wt.% Nd	DMEM, 37°C	VSMC, IC	5 days	81.1	—	[369]
Mg–2Zn–0.5Y–0.5Nd	0.5 wt.% Y + 0.5 wt.% Nd	PBS + Diluted blood, 37°C	—	—	—	5.1	[367]
Mg–2Zn–0.5Y–0.5Nd	0.5 wt.% Y + 0.5 wt.% Nd	DMEM, 37°C	HUVEC, DC	1 day	104.7	—	[367]
Mg–2Zn–0.5Y–0.5Nd	0.5 wt.% Y + 0.5 wt.% Nd	DMEM, 37°C	HUVEC, DC	2 days	52.1	—	[367]
Mg–2Zn–0.5Y–0.5Nd	0.5 wt.% Y + 0.5 wt.% Nd	DMEM, 37°C	HUVEC, DC	3 days	32.3	—	[367]
Mg–2Zn–0.5Y–0.5Nd	0.5 wt.% Y + 0.5 wt.% Nd	DMEM, 37°C	HAMSC, DC	1 day	78.2	—	[367]
Mg–2Zn–0.5Y–0.5Nd	0.5 wt.% Y + 0.5 wt.% Nd	DMEM, 37°C	HAMSC, DC	2 days	71.2	—	[367]
Mg–2Zn–0.5Y–0.5Nd	0.5 wt.% Y + 0.5 wt.% Nd	DMEM, 37°C	HAMSC, DC	3 days	78.8	—	[367]

Material	Composition	Medium	Cell/Test	Time	Value 1	Value 2	Ref
Mg–2Zn–0.5Y–0.5Nd (extruded)	0.5 wt.% Y + 0.5 wt.% Nd	PBS + Diluted blood, 37°C	—	—	—	16.4	[368]
Mg–2Zn–0.5Y–0.5Nd (extruded)	0.5 wt.% Y + 0.5 wt.% Nd	DMEM, 37°C	HUVEC, DC	1 day	63.8	—	[368]
Mg–2Zn–0.5Y–0.5Nd (extruded)	0.5 wt.% Y + 0.5 wt.% Nd	DMEM, 37°C	HUVEC, DC	2 days	73.3	—	[368]
Mg–2Zn–0.5Y–0.5Nd (extruded)	0.5 wt.% Y + 0.5 wt.% Nd	DMEM, 37°C	HUVEC, DC	3 days	67.9	—	[368]
Mg–2Zn–0.5Y–0.5Nd (extruded)	0.5 wt.% Y + 0.5 wt.% Nd	DMEM, 37°C	HAMSC, DC	1 day	73.0	—	[368]
Mg–2Zn–0.5Y–0.5Nd (extruded)	0.5 wt.% Y + 0.5 wt.% Nd	DMEM, 37°C	HAMSC, DC	2 days	71.4	—	[368]
Mg–2Zn–0.5Y–0.5Nd (extruded)	0.5 wt.% Y + 0.5 wt.% Nd	DMEM, 37°C	HAMSC, DC	3 days	60.5	—	[368]
Mg–Y–Nd (extruded)	—	α-MEM, 37°C	L929, IC	1 day	96.9	—	[370]
Mg–Y–Nd (extruded)	—	α-MEM, 37°C	MC3T3-E1, IC	1 day	101.9	—	[370]
Mg–Y–Nd–Zr (extruded)	—	α-MEM, 37°C	L929, IC	1 day	109.3	—	[370]
Mg–Y–Nd–Zr (extruded)	—	α-MEM, 37°C	MC3T3-E1, IC	1 day	113.4	—	[370]

* All the Mg–Zn–RE alloys are cast, unless otherwise specified.

the strength and, above all, the ductility. The elongation to failure of as-cast pure Mg decreases from 7.31% to 3.05% and 2.1% by adding 2 and 4 wt.% Ca, respectively. This is due to the continuous distribution of brittle Mg_2Ca along grain boundaries or interdendritic region. However, full refinement of the Mg_2Ca phase into small isolated particles and their uniform dispersion over the matrix are reported to suppress the growth of large cracks or voids in the early stage of plastic deformation, producing good tensile ductility and an increase in the tensile strength by increasing the Ca content. This can be achieved through extrusion, rolling and other plastic deformation techniques. Miyazaki et al. [379] reported an increase in the tensile strength of extruded Mg–Ca alloys by adding up to 5 wt.% Ca, and Jeong and Kim [2] reported the elongation to failure of Mg–1Ca to increase from 4.1% to 19.5% after extrusion. In addition, they reported extruded Mg–3Ca to have an elongation to failure of 7.3%, far higher than the 0.26% of as-cast Mg–3Ca obtained in ref. [380], in agreement with ref. [233]. The improvement in the ductility is also due to a weaker texture. In fact, previous studies on Ca containing Mg alloys demonstrated that the DRXed grains usually represent a relatively weak texture. The grains nucleated by the particle-stimulated nucleation mechanism are proven to be random in orientation, which leads to the weak texture [381,382], activating basal and prismatic slips [383]. Moreover, increasing the extrusion temperature leads to a lower texture intensity [384]. Further examples of the improvement in the ductility provided by the thermo-mechanical process compared to the as-cast Mg–Ca alloys are gathered in Table 3.37.

3.5.2   Mg–Ca Alloys: Corrosion Resistance

According to the Mg–Ca phase diagram, the maximum solubility of Ca in Mg at 789.5 K is 1.34 wt.% [385]. When the solubility limit of Ca is exceeded, the formation of the intermetallic phase Mg_2Ca takes place. Differently from the other intermetallic phases, Mg_2Ca has been found to be anodic with respect to the α-Mg matrix. Cha et al. [232] reported the matrix, characterized by an open circuit potential (OCP) of −1.58 V, is nobler than the intermetallic phase Mg_2Ca (OCP = −1.87 V). The influence of this second phase on the corrosion resistance is quite well understood. Despite the work of Kim et al. [386] where they reported the corrosion resistance of as-cast Mg–xCa (x = 0, 0.5, 1, 5 wt.%) to increase with the increase in Ca content. The authors attributed this improvement to the barrier effect provided by the accumulation of precipitates at the grain boundaries, where other studies available reported the opposite trend. Harandi et al. [387] reported that the corrosion resistance is enhanced by the Ca addition when this is below its solubility limit, but, when the Ca content is too high, the Mg_2Ca phases play a harmful role to the corrosion resistance due to the onset of the galvanic cell between the α-Mg matrix and the second phase. In particular, Jeong and Kim [2] reported that if the volume fraction of Mg_2Ca phase is high enough to surround the α-Mg grain (when Ca content is higher than 2 wt.%), the corrosion initiated at the Mg_2Ca phase in the eutectic structure penetrates speedily and continuously

TABLE 3.37

Summary of the mechanical properties of Mg–Ca alloys.*

Alloy	Additional elements (wt.%)	Yield strength (MPa)	UTS (MPa)	Elongation to failure (%)	Compressive yield strength (MPa)	Ultimate compressive strength (MPa)	Ref.
Mg–0.5Ca	—	—	—	—	—	166.5	[378]
Mg–1Ca	—	—	—	—	—	178.9	[378]
Mg–2Ca	—	—	—	—	—	184.7	[378]
Mg–5Ca	—	—	—	—	—	188.2	[378]
Mg–10Ca	—	—	—	—	—	190.0	[378]
Mg–15Ca	—	—	—	—	—	207.8	[378]
Mg–20Ca	—	—	—	—	—	291.1	[378]
Mg–1Ca–1Y	1 Y	—	—	—	—	159.9	[378]
Mg–2Ca (RS + cold pressed + extruded)	—	—	380	7.3	—	—	[379]
Mg–2Ca (RS + cold pressed + extruded)	—	—	458	0.9	—	—	[379]
Mg–5Ca–5Zn (RS + cold pressed + extruded)	5 Zn	—	483	2.0	—	—	[379]
Mg–5Ca–3Si (RS + cold pressed + extruded)	3 Si	—	260	6.4	—	—	[379]
Mg–0.19Ca–0.25Si	0.25 Si	—	—	—	45.5	—	[376]
Mg–0.52Ca–0.5Si	0.5 Si	—	—	—	65.3	—	[376]
Mg–0.5Ca–0.55Si	0.55 Si	—	—	—	68.1	—	[376]
Mg–0.29Ca–0.19Si	0.19 Si	—	—	—	45.3	—	[376]
Mg–0.18Ca–0.15Si	0.15 Si	—	—	—	42.7	—	[376]
Mg–0.20Ca–0.15Si	0.15 Si	—	—	—	49.1	—	[376]
Mg–1.34Ca–0.41Si	0.41 Si	—	—	—	74.1	—	[376]
Mg–1.43Ca–0.16Si	0.16 Si	—	—	—	70.9	—	[376]
Mg–1.7Ca–0.1Si	0.1 Si	—	—	—	84.0	—	[376]
Mg–2Ca (extruded)	—	205.3	262.0	14.6	—	—	[5]

(*Continued*)

TABLE 3.37 (Cont.)

Alloy	Additional elements (wt.%)	Yield strength (MPa)	UTS (MPa)	Elongation to failure (%)	Compressive yield strength (MPa)	Ultimate compressive strength (MPa)	Ref.
Mg–3Ca (extruded)	—	249.4	284.4	7.4	—	—	[5]
Mg–2Ca (extruded + HRDSR)	—	—	184.6	—	—	—	[5]
Mg–3Ca (extruded + HRDSR)	—	—	229.4	—	—	—	[5]
Mg–2Ca (extruded + HRDSR + annealed)	—	136.2	155.7	2.4	—	—	[5]
Mg–3Ca (extruded HRDSR + annealed)	—	162.1	192.1	1.9	—	—	[5]
Mg–1Ca (homogenized + indirect extruded)	—	280	316.2	10.9	—	—	[384]
Mg–1Ca	—	41.2	72.2	1.9	—	—	[233]
Mg–2Ca	—	36.9	52.7	1.1	—	—	[233]
Mg–3Ca	—	13.6	44.0	0.6	—	—	[233]
Mg–1Ca (rolled)	—	123.6	167.3	3.1	—	—	[233]
Mg–1Ca (extruded)	—	135.7	240.1	10.6	—	—	[233]
Mg–3Ca	—	110	118	0.26	—	—	[380]
Mg–3Ca–2Zn	2 Zn	117	145	0.57	—	—	[380]
Mg–0.4Ca	—	42.3	78.2	10.8	—	—	[2]
Mg–1Ca	—	81.3	98.8	4.1	—	—	[2]
Mg–0.4Ca (homogenized + indirectly extruded)	—	165.6	234.1	34.0	—	—	[2]
Mg–1Ca (homogenized + indirectly extruded)	—	185.1	239.3	19.5	—	—	[2]
Mg–2Ca (homogenized + indirectly extruded)	—	204.7	252.8	14.6	—	—	[2]

(Continued)

TABLE 3.37 (Cont.)

Alloy	Additional elements (wt.%)	Yield strength (MPa)	UTS (MPa)	Elongation to failure (%)	Compressive yield strength (MPa)	Ultimate compressive strength (MPa)	Ref.
Mg–3Ca (homogenized + indirectly extruded)	—	248.9	273.8	7.3	—	—	[2]
Mg–2Ca	—	47.3	115.2	3.1	—	235.7	[235]
Mg–4Ca	—	34.5	77.4	2.1	—	250.7	[235]
Mg–2Ca–0.5Mn–2Zn	0.5 Mn + 2 Zn	8.3	168.5	7.8	—	336.1	[235]
Mg–2Ca–0.5Mn–4Zn	0.5 Mn + 4 Zn	83.1	189.2	8.7	—	343.6	[235]
Mg–2Ca–0.5Mn–7Zn	0.5 Mn + 7 Zn	45.4	140.7	4.2	—	362.8	[235]

* All the Mg–Ca alloys are cast, unless otherwise specified. Additional information can be found on www.routledge.com/9780367429454.

along the grain boundaries, promoting deep pitting in the matrix as well as its undercutting, leading to a fast loss of large weight (Figure 3.14a).

They also reported a high amount of Ca to improve the corrosion resistance by means of break-up and dispersion of the eutectic Mg_2Ca phase via a thermo-mechanical processes, which is in agreement with ref. [5]. They also considered that when the Mg_2Ca phase is discontinuous at the grain boundaries, the corrosion front does not penetrate the grain boundaries of the matrix, thus interrupting the propagation of corrosion (Figure 3.14b). Further examples of the effect of Ca and of the thermo-mechanical processes on the corrosion resistance are gathered in Tables 3.38 and 3.39.

3.5.3 MG–CA ALLOYS: BIOCOMPATIBILITY

When Mg–Ca alloys were started to be studied, a high biocompatibility was expected. Mg is one of the most abundant element in the human body with significant functional roles in soft biological systems and bone tissues [388–391], having stimulatory effects on the growth of new bone tissues [14,224,392,393]. In addition, Ca is a very important element in the human body and 99% or more of the Ca is deposited in bones and the remainder is associated with nerve conduction, muscle contraction, hormone release and cell signaling [394]. The results in terms of *in vitro* and *in vivo* biocompatibility are promising (Table 3.40). With the exception of as-cast Mg–2Ca and Mg–3Ca, where the corrosion rate is too high and leads to an increase in the pH value of the cell-culturing solution, deleterious for cell adhesion, growth and proliferation [378,395], Mg/Ca ions were reported to effectively induce the

a)

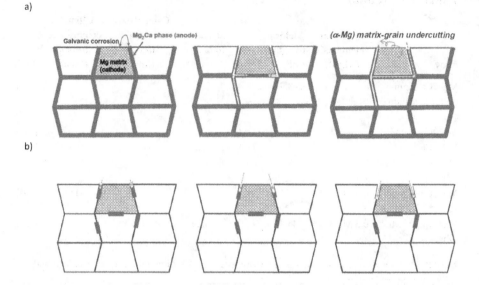

b)

FIGURE 3.14 Schematic illustration of the role of Mg₂Ca phase continuously (a) and discontinuously (b) distributed in the corrosion mechanisms [2].

growth of osteoblast cells, even at high extract concentration [396]. Li et al. [233] proposed a biocorrosion model at the alloy/aqueous solution interface to understand the corrosion processes and the subsequent hydroxyapatite formation to explain the active bone formation of Mg–Ca alloys. Moreover, Cho et al. [397] implanted Mg–Ca–Zn screws on New Zealand white rabbits and they reported that the chemical composition and degradation rate of the Mg–Ca–Zn alloy bone screw do not induce a negative effect to the body. In addition, compared to self-reinforced poly-L-lactide (SR-PLLA) screws, the Mg–Ca–Zn alloy bone screws were characterized by lower inflammatory scores. These results agree with those obtained by Berglund et al. [398], who compared the *in vivo* biocompatibility of Mg–1Ca–0.5Sr alloy and PLLA pins on Sprague–Dawley rats. In addition, He et al. [399] reported that the results of direct antibacterial assays on Mg–1Ca–0.5Sr alloy show not only the inhibition of bacteria's growth, but also a decreased number of adhered bacteria on the alloy surfaces, while the adhered bacteria on the Ti–6Al–4V surface increases with incubation time.

3.6 MG–RE ALLOYS

The development of high-strength Mg–RE alloys in industrial applications has attracted the interest of biomedical researchers who are focused on increasing the mechanical properties of Mg alloys. Yttrium (Y) and Gadolinium (Gd)

TABLE 3.38

Summary of the DC polarization curve results for Mg–Ca alloys.*

Alloy	Additional elements	Corrosive environment	E_{corr} (V)	i_{corr} ($\mu A/cm^2$)	Ref.
Mg–0.5Ca	—	Tas–SBF, 36.5°C	−1.58	—	[386]
Mg–1Ca	—	Tas–SBF, 36.5°C	−1.45	—	[386]
Mg–1.5Ca	—	Tas–SBF, 36.5°C	−1.39	—	[386]
Mg–3Ca (extruded + annealed)	—	Hank's, 37°C	−1.58	8.9	[5]
Mg–2Ca (extruded + HRDSR)	—	Hank's, 37°C	−1.52	39.8	[5]
Mg–3Ca (extruded + HRDSR)	—	Hank's, 37°C	−1.59	16.1	[5]
Mg–2Ca (extruded + HRDSR + annealed)	—	Hank's, 37°C	−1.59	6.8	[5]
Mg–3Ca (extruded + HRDSR + annealed)	—	Hank's, 37°C	−1.62	3.9	[5]
Mg–0.7Ca (homogenized)	—	SBF, 37°C	−1.90	197	[387]
Mg–1Ca (homogenized)	—	SBF, 37°C	−1.96	224	[387]
Mg–2Ca (homogenized)	—	SBF, 37°C	−1.99	312	[387]
Mg–3Ca (homogenized)	—	SBF, 37°C	−2.05	395	[387]
Mg–4Ca (homogenized)	—	SBF, 37°C	−2.06	470	[387]
Mg–0.4Ca	—	Hank's, 37°C	−1.67	7.0	[2]
Mg–1Ca	—	Hank's, 37°C	−1.69	9.8	[2]
Mg–2Ca	—	Hank's, 37°C	−1.64	10.5	[2]
Mg–3Ca	—	Hank's, 37°C	−1.60	25.6	[2]
Mg–0.4Ca (homogenized + indirect extruded)	—	Hank's, 37°C	−1.62	7.8	[2]
Mg–1Ca (homogenized + indirect extruded)	—	Hank's, 37°C	−1.65	8.45	[2]
Mg–2Ca (homogenized + indirect extruded)	—	Hank's, 37°C	−1.59	8.5	[2]
Mg–3Ca (homogenized + indirect extruded)	—	Hank's, 37°C	−1.58	10.7	[2]
Mg–1.2Ca (extruded)	—	Hank's, 37°C	−1.55	63.0	[60]
Mg–1.2Ca (extruded)	—	DMEM, 37°C	−1.52	24.1	[60]
Mg–1.2Ca (extruded)	—	DMEM + 10% FBS, 37°C	−1.54	39.5	[60]
Mg–2Ca	—	SBF, 37°C	−1.99	301,9	[235]
Mg–4Ca	—	SBF, 37°C	−2.05	395.7	[235]
Mg–0.5Ca	—	SBF, 37°C	−1.88	186	[257]
Mg–0.5Ca–0.5Zn	0.5 wt.% Zn	SBF, 37°C	−1.71	189	[257]
Mg–1Ca	—	SBF, 37°C	−1.89	—	[233]
Mg–2Ca	—	SBF, 37°C	−1.88	—	[233]
Mg–3Ca	—	SBF, 37°C	−1.92	—	[233]
Mg–1Ca (extruded)	—	SBF, 37°C	−1.73	—	[233]
Mg–1Ca (rolled)	—	SBF, 37°C	−179	—	[233]

* All the Mg–Ca alloys are cast, unless otherwise specified.

TABLE 3.39

Summary of the corrosion rates for Mg–Ca alloys.*

Alloy	Additional elements	Corrosive environment	Immersion time	Corrosion rate (mm/ year)	Procedure	Ref.
Mg–0.5Ca	—	Tas–SBF, 36.5°C	14 days	4.4	H	[386]
Mg–1Ca	—	Tas–SBF, 36.5°C	14 days	3.7	H	[386]
Mg–1.5Ca	—	Tas–SBF, 36.5°C	14 days	3.3	H	[386]
Mg–3Ca (extruded + annealed)	—	Hank's, 37°C	1 day	0.8	WL	[5]
Mg–3Ca (extruded + annealed)	—	Hank's, 37°C	7 days	0.95	WL	[5]
Mg–2Ca (extruded + HRDSR)	—	Hank's, 37°C	1 day	2.3	WL	[5]
Mg–2Ca (extruded + HRDSR)	—	Hank's, 37°C	7 days	2.5	WL	[5]
Mg–3Ca (extruded + HRDSR)	—	Hank's, 37°C	1 day	4.9	WL	[5]
Mg–3Ca (extruded + HRDSR)	—	Hank's, 37°C	7 days	4.2	WL	[5]
Mg–2Ca (extruded + HRDSR + annealed)	—	Hank's, 37°C	1 day	0.9	WL	[5]
Mg–2Ca (extruded + HRDSR + annealed)	—	Hank's, 37°C	7 days	0.08	WL	[5]
Mg–3Ca (extruded + HRDSR + annealed)	—	Hank's, 37°C	1 day	1.2	WL	[5]
Mg–3Ca (extruded + HRDSR + annealed)	—	Hank's, 37°C	7 days	0.15	WL	[5]
Mg–0.8Ca (homogenized + extruded)	—	Rabbit (in vivo)	12 weeks	0.3	WL	[127]
Mg–0.8Ca (homogenized + extruded)	—	Rabbit (in vivo)	24 weeks	0.46	WL	[127]
Mg–0.4Ca	—	Hank's, 37°C	1 day	0.13	WL	[2]
Mg–0.4Ca	—	Hank's, 37°C	3 days	0.16	WL	[2]
Mg–0.4Ca	—	Hank's, 37°C	7 days	0.36	WL	[2]
Mg–1Ca	—	Hank's, 37°C	1 day	0.87	WL	[2]
Mg–1Ca	—	Hank's, 37°C	3 days	0.75	WL	[2]
Mg–1Ca	—	Hank's, 37°C	7 days	1.69	WL	[2]
Mg–2Ca	—	Hank's, 37°C	1 day	5.31	WL	[2]
Mg–2Ca	—	Hank's, 37°C	3 days	5.71	WL	[2]

(Continued)

TABLE 3.39 (Cont.)

Alloy	Additional elements	Corrosive environment	Immersion time	Corrosion rate (mm/ year)	Procedure	Ref.
Mg–2Ca	—	Hank's, 37°C	7 days	10.88	WL	[2]
Mg–3Ca	—	Hank's, 37°C	1 day	22.31	WL	[2]
Mg–3Ca	—	Hank's, 37°C	3 days	19.11	WL	[2]
Mg–3Ca	—	Hank's, 37°C	7 days	21.95	WL	[2]
Mg–0.4Ca (homogenized + indirect extruded)	—	Hank's, 37°C	1 day	0.23	WL	[2]
Mg–0.4Ca (homogenized + indirect extruded)	—	Hank's, 37°C	3 days	0.098	WL	[2]
Mg–0.4Ca (homogenized + indirect extruded)	—	Hank's, 37°C	7 days	0.32	WL	[2]
Mg–1Ca (homogenized + indirect extruded)	—	Hank's, 37°C	1 day	0.30	WL	[2]
Mg–1Ca (homogenized + indirect extruded)	—	Hank's, 37°C	3 days	0.12	WL	[2]
Mg–1Ca (homogenized + indirect extruded)	—	Hank's, 37°C	7 days	0.29	WL	[2]
Mg–2Ca (homogenized + indirect extruded)	—	Hank's, 37°C	1 day	0.41	WL	[2]
Mg–2Ca (homogenized + indirect extruded)	—	Hank's, 37°C	3 days	0.16	WL	[2]
Mg–2Ca (homogenized + indirect extruded)	—	Hank's, 37°C	7 days	0.69	WL	[2]
Mg–3Ca (homogenized + indirect extruded)	—	Hank's, 37°C	1 day	0.68	WL	[2]
Mg–3Ca (homogenized + indirect extruded)	—	Hank's, 37°C	3 days	0.35	WL	[2]
Mg–3Ca (homogenized + indirect extruded)	—	Hank's, 37°C	7 days	0.51	WL	[2]
Mg–2Ca	—	SBF, 37°C	10 days	6.4	H	[235]

(Continued)

TABLE 3.39 (Cont.)

Alloy	Additional elements	Corrosive environment	Immersion time	Corrosion rate (mm/ year)	Procedure	Ref.
Mg–4Ca	—	SBF, 37°C	10 days	14.4	H	[235]
Mg–1Ca	—	SBF, 37°C	10 days	7.3	H	[233]
Mg–1Ca (extruded)	—	SBF, 37°C	10 days	2.1	H	[233]
Mg–1Ca	—	Rabbit (in vivo)	3 months	2.3	WL	[233]

* All the Mg–Ca alloys are cast, unless otherwise specified. Additional information can be found on www.routledge.com/9780367429454.

are, in fact, reported to provide both solid solution strengthening and precipitate strengthening [400–405]. WE43 alloys (Mg–4Y–3RE) have been intensively investigated. In addition, the mechanical properties can be further tuned by adding Zn due to the formation of LPSO structures (see Section 3.4.6.1). Recently, magnesium alloys with LPSO structure have attracted much attention due to their excellent mechanical properties. Mg–RE–Zn alloys with LPSO phase can be classified into two types [1]: for type I alloys such as Mg–Y–Zn, 18R LPSO phase formed in as-cast alloy changes to 14H after solution treatment; for type II alloys such as Mg–Gd–Zn, no LPSO phase is formed in as-cast alloy, but 14H LPSO phase is precipitated during solution treatment. The addition of RE element, in particular Gd and Y, is also reported to be beneficial for the corrosion behavior. The positive effect of RE elements is attributed to two mechanisms, the "scavenger effect" (interaction of RE with impurities such as Fe, Ni and Co that leads to less noble intermetallic phases) and the formation of a protective passivation layer that improves the stability of the hydroxide layer against Cl^-. Mg–RE alloys have then undergone animal and clinical trials. The Magic stent (Biotronik Inc., Germany) is a WE43 stent that in 2010 was reported to be the only magnesium stent to have undergone clinical trials, showing safety and similar scaffolding properties as stainless steel, however, it degraded too fast [406]. Developments on the stents produced by Biotronik Inc. is reviewed in ref. [407]. Magnezix® compression screw (Sintellix AG, Hannover, Germany) is classified as Mg–Y–RE–Zr alloy according to DIN EN 1753 and has been studied on a trauma application where the implant was used for an intra-articular fracture fixation in the elbow [408].

3.6.1 Mg–RE Alloys: Mechanical Properties

Using Gd and Y as primary alloying elements in Mg–RE alloys generally leads to higher mechanical properties because of their high capability of providing solid solution strengthening and precipitate strengthening. In particular, they are ideal alloying elements for precipitation hardening: the equilibrium solubility of Gd and Y in Mg decreases exponentially from 23.49 wt.% and

TABLE 3.40
Summary of the biocompatibility of Mg–Ca alloys.*

Alloy	Additional elements	Test environment	Cell type, procedure	Time of culture	Result (%)	Hemolysis (%)	Ref.
			Cell viability				
Mg–1Ca–0.5Sr (homogenized + extruded)	0.5 wt.% Sr	α-MEM, 37°C	MC3T3-E1, IC	2 days	100.0	—	[399]
Mg–1Ca–0.5Sr (homogenized + extruded)	0.5 wt.% Sr	α-MEM, 37°C	MC3T3-E1, IC	4 days	115.4	—	[399]
Mg–1Ca–0.5Sr (homogenized + extruded)	0.5 wt.% Sr	α-MEM, 37°C	MC3T3-E1, IC	6 days	80.5	—	[399]
Mg–30Ca	—	DMEM, 37°C	L-929, IC	2 days	100.8	—	[396]
Mg–30Ca	—	DMEM, 37°C	L-929, IC	4 days	107.3	—	[396]
Mg–30Ca	—	DMEM, 37°C	L-929, IC	7 days	104.7	—	[396]
Mg–3Ca	—	DMEM, 37°C	L-929, IC	1 day	64.1	—	[395]
Mg–3Ca	—	DMEM, 37°C	L-929, IC	2 days	57.1	—	[395]
Mg–3Ca	—	DMEM, 37°C	L-929, IC	4 days	55.8	—	[395]
Mg–3Ca (rapid solidified)	—	DMEM, 37°C	L-929, IC	1 day	96.9	—	[395]
Mg–3Ca (rapid solidified)	—	DMEM, 37°C	L-929, IC	2 days	102.6	—	[395]
Mg–3Ca (rapid solidified)	—	DMEM, 37°C	L-929, IC	4 days	107.1	—	[395]
Mg–1Ca	—	DMEM, 37°C	L-929, IC	2 days	101.3	—	[233]
Mg–1Ca	—	DMEM, 37°C	L-929, IC	4 days	102.3	—	[233]
Mg–1Ca	—	DMEM, 37°C	L-929, IC	7 days	105.7	—	[233]
Mg–0.5Ca	—	MMEM, 37°C	SaOS2, IC	1 day	93.9	—	[378]
Mg–1Ca	—	MMEM, 37°C	SaOS2, IC	1 day	93.4	—	[378]
Mg–2Ca	—	MMEM, 37°C	SaOS2, IC	1 day	59.9	—	[378]
Mg–1Ca–1Y	1 wt.% Y	MMEM, 37°C	SaOS2, IC	1 day	93.8	—	[378]

* All the Mg–Ca alloys are cast, unless otherwise specified. Additional information can be found on www.routledge.com/9780367429454.

12.47 wt.% at the eutectic temperature (548°C and 566°C, respectively) to 3.82 wt.% and 2.69 wt.% at 200°C [409,410]. During isothermal aging treatment, the precipitation sequence can be concluded as follows: Mg supersaturated solid solution $(Mg)S.S.S.S-\beta''-\beta'-\beta_1-\beta_e$ and their effects on the mechanical properties are reported in ref. [411]. In addition, adding Ag was reported to increase the mechanical properties of both Mg–Gd and Mg–Y alloys due to a significantly enhanced age-hardening response associated with the formation of nano-scale basal precipitates on (0001)α [412]. Zhu et al. reported the addition of 0.4 wt.% Ag leads to an increase in UTS of 11% for peak-aged Mg–2.8Y–0.1Zr and of 41% in the case of peak-aged Mg–2.8Gd–0.1Zr [413]. Moreover, the addition of Zn to Mg–RE has been investigated due to the high mechanical properties achievable due to the presence of LPSO structures, that is 18R LPSO phase in as-cast Mg–Y–Zn and 14H LPSO phase in solution treated Mg–Y–Zn and Mg–Gd–Zn. In fact, the addition of 0.2 wt.% Zn in WE43 leads to achieve a tensile strength and a yield strength of 345 and 196 MPa, respectively, much higher than its typical values (250 MPa and 162 MPa, respectively) [414]. Researchers have then tried to improve the mechanical properties, in particular the ductility, of Mg–RE alloys using thermo-mechanical processes such as extrusion and rolling (sometimes also in combination with thermal treatments) and SPD techniques. Panigrahi et al. [415] reported the forging treatment to increase the elongation to failure on artificially aged WE43. In particular, the thermo-mechanical processes combined with RE-containing alloys lead to a weak and random texture that increases the ductility [352,416]. Wu et al. [417] reported that after hot extrusion and annealing, the texture of the pure magnesium is quite different from those of the Mg–Y alloys, in particular increasing the yttrium content leads to a weaker and more random texture (Figure 3.15).

The difference in texture is reported to be due to the introduction of non-basal slip. In the case of extrusion at high temperature, pyramidal slip in the Mg–Y alloys will be activated because of the reduction of the axis ratio and the decrease of the critical resolved shear stress. These results are in agreement with those reported by Cottam et al. [418]. They reported that after high temperature extrusion, mechanical twins can be found in pure magnesium, whereas by increasing the yttrium content (0.23 wt.% and 0.84 wt.%) the evidence of twinning disappears in favor of non-basal <c+a> slip. In addition, a weak texture due to the activation of prismatic and <c+a> pyramidal slip was observed in ECAP WE43. Further data about the improvement of ductility due to activation of nonbasal slips plane is reported in Table 3.41, and the possible reasons behind this phenomena are gathered in ref. [417].

3.6.2 MG–RE ALLOYS: CORROSION RESISTANCE

The corrosion behaviors of Mg–RE alloys have been widely studied but their effects are still debated. The results are dependent on their concentration, on the combination of RE elements and on the thermo-mechanical processes which the alloys were subjected to. The "scavenger effect" and their passivation properties are in fact contrasted by the galvanic effect of the second phases. Therefore, while Hort et al. [419] reported the as-cast Mg–10Gd alloy to be characterized by the lowest corrosion rate among Mg–Gd alloy with a Gd concentration ranging from 2 wt.% to 15 wt.%, Chang et al. [420]

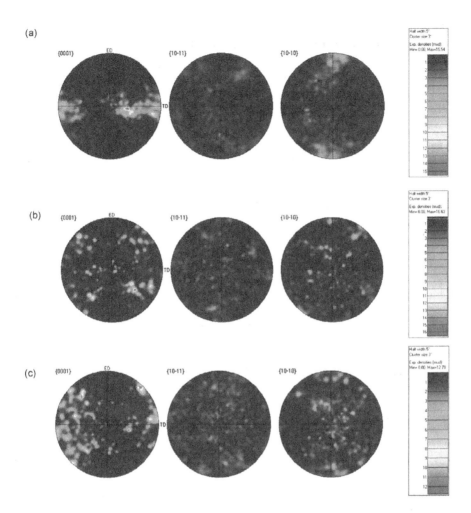

FIGURE 3.15 Pole figures of extruded and annealed pure Mg (a), Mg–2Y (b) and Mg–4Y (c). Reprinted with permission from Elsevier [417].

reported that T6 heat-treated Mg−10Gd−3Y−0.4Zr was characterized by the highest corrosion rate when compared with the same alloys characterized by a concentration of Gd ranging from 6 wt.% to 12 wt.%. The same controversial results were obtained for Mg–Nd. Tanenaka et al. [421] reported a positive effect of 1 wt.% Nd addition on Mg, whereas Birbilis et al. [422] reported the addition of Nd in a range of 0.5–3.5 wt.% to decrease the corrosion resistance. Again, the same happened when Yttrium was considered. Liu et al. [423] showed that increasing the amount of yttrium led to a faster hydrogen evolution (Figure 3.16) due to microgalvanic corrosion, while Chou

TABLE 3.41

Summary of the mechanical properties of Mg–RE alloys.*

Alloy	Additional elements (wt.%)	Yield strength (MPa)	UTS (MPa)	Elongation to failure (%)	Compressive yield strength (MPa)	Ultimate compressive strength (MPa)	Ref.
$Mg_{97.1}Gd_{2.8}Zr_{0.1}$ (solution treated)	0.1 at.% Zr	166	283	12.2	—	—	[413]
$Mg_{97.1}Gd_{2.8}Ag_{0.4}$ $Zr_{0.1}$ (solution treated)	0.1 at.% Zr + 0.4 at.% Ag	155	296	19.4	—	—	[413]
$Mg_{97.1}Gd_{2.8}Zr_{0.1}$ (solution treated + aged)	0.1 at.% Zr	246	293	0.4	—	—	[413]
$Mg_{97.1}Gd_{2.8}Ag_{0.4}$ $Zr_{0.1}$ (solution treated + aged)	0.1 at.% Zr + 0.4 at.% Ag	271	414	2.7	—	—	[413]
$Mg_{97.1}Y_{2.8}Zr_{0.1}$ (solution treated)	0.1 at.% Zr	147	222	3.7	—	—	[413]
$Mg_{97.1}Y_{2.8}Ag_{0.4}$ $Zr_{0.1}$ (solution treated)	0.1 at.% Zr + 0.4 at.% Ag	140	260	19.3	—	—	[413]
$Mg_{97.1}Y_{2.8}Zr_{0.1}$ (solution treated + aged)	0.1 at.% Zr	238	278	0.4	—	—	[413]
$Mg_{97.1}Y_{2.8}Ag_{0.4}$ $Zr_{0.1}$ (solution treated + aged)	0.1 at.% Zr + 0.4 at.% Ag	217	309	2.4	—	—	[413]
WE43 (T6)	0.2 Zn	240	298	4.6	—	—	[414]
WE43	0.2 Zn	162	205	5.0	—	—	[414]
WE43 (T6)	0.2 Zn	225	304	6.4	—	—	[414]
WE43 (rolled)	—	185	261	31	—	—	[415]
WE43 (rolled + T5)	—	270	348	16	—	—	[415]
WE43 (rolled + forged)	—	263	311	23	—	—	[415]
WE43 (rolled + forged + aged)	—	344	388	23	—	—	[415]
Mg–2Y (homogenized + extruded + annealed)	—	92	189	21	—	—	[417]
Mg–2Y (homogenized + extruded + annealed)	—	87	177	30	—	—	[417]

(Continued)

TABLE 3.41 (Cont.)

Alloy	Additional elements (wt.%)	Yield strength (MPa)	UTS (MPa)	Elongation to failure (%)	Compressive yield strength (MPa)	Ultimate compressive strength (MPa)	Ref.
WE43	—	149	206	4.0	—	—	[411]
WE43 (as quenched)	—	147	227	8.0	—	—	[411]
WE43 (T6)	—	202	282	1.0	—	—	[411]
WE43	—	161	282.8	18.3	—	—	[406]
WE43 (extruded)	—	201	290	17	—	—	[118]
WE54 (homogenized + extruded + T4 + T6)	—	172.6	—	—	156.4	—	[350]

* All the Mg–RE alloys are cast, unless otherwise specified. Additional information can be found on www.routledge.com/9780367429454.

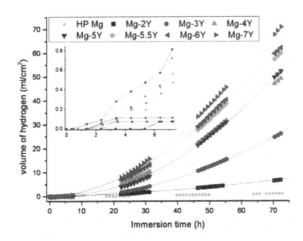

FIGURE 3.16 Hydrogen evolution of Mg–Y alloys with various yttrium content immersed in 0.1 M NaCl. Reprinted with permission from Elsevier [423].

et al. [424] reported the addition of Y on Mg–Y–Ca–Zr alloys to be beneficial for the corrosion resistance.

Discrepant results are also reported for thermal treatment. In fact, while T4 heat treatment has been reported to have positive effects on the corrosion resistance of Mg–4Nd, harmful effects were reported for Mg–5Gd and no marked changes for WX41 [424,425], and further results are reported in Tables 3.42 and 3.43.

TABLE 3.42

Summary of the DC polarization curve results for Mg–RE alloys.*

Alloy	Additional elements	Corrosive environment	E_{corr} (V)	i_{corr} ($\mu A/cm^2$)	Ref.
Mg–1Y–0.6Ca–0.4Zr	0.6 wt.% Ca + 0.4 wt.% Zr	DMEM + 10% FBS, 37°C	−1.73	14.9	[424]
Mg–1Y–0.6Ca–0.4Zr (T4)	0.6 wt.% Ca + 0.4 wt.% Zr	DMEM + 10% FBS, 37°C	−1.58	24.9	[424]
Mg–4Y–0.6Ca–0.4Zr	0.6 wt.% Ca + 0.4 wt.% Zr	DMEM + 10% FBS, 37°C	−1.59	4.9	[424]
Mg–4Y–0.6Ca–0.4Zr (T4)	0.6 wt.% Ca + 0.4 wt.% Zr	DMEM + 10% FBS, 37°C	−1.54	5.2	[424]
Mg–1Gd	—	9 g/l, 37°C	−1.74	27	[425]
Mg–3Gd	—	9 g/l, 37°C	−1.74	23	[425]
Mg–5Gd	—	9 g/l, 37°C	−1.74	13	[425]
Mg–5Gd (T4)	—	9 g/l, 37°C	−1.65	33	[425]
Mg–1Nd	—	9 g/l, 37°C	−1.63	11	[425]
Mg–4Nd	—	9 g/l, 37°C	−1.62	9	[425]
Mg–4Nd (T4)	—	9 g/l, 37°C	−1.84	18	[425]
Mg–9Nd	—	9 g/l, 37°C	−1.56	20	[425]
Mg–1Gd–1Y	—	9 g/l, 37°C	−1.64	24	[425]
Mg–3Gd–1Y	—	9 g/l, 37°C	−1.68	25	[425]
Mg–5Gd–1Y	—	9 g/l, 37°C	−1.69	14	[425]

* All the Mg–RE alloys are cast, unless otherwise specified. Additional information can be found on www.routledge.com/9780367429454.

3.6.3　Mg–RE Alloys: Biocompatibility

As reported in Sections 3.4.6 and 3.4.6.3, the biocompatibility of RE elements is still debated. For example, high amounts of yttrium (above its daily intake of 4.2 mg/day) can cause increases of blood eosinocyte, decreases of body weight and eosinophil infiltration in the submucosa. However, another study indicated that Mg alloys containing small amounts of Y and RE elements would be appropriate for biomedical applications since yttrium has been shown to be nontoxic and nonhepatotoxic in longevity studies [151,426]. Feyerabend et al. [363] evaluated the viability and inflammatory effects of different RE elements (Figure 3.17) and reported Dy and Gd to be characterized by a higher viability than yttrium.

The majority of the studies on the biocompatibility of Mg–RE alloys assessed WE43 alloy with contradicting results. Naujokat et al. [427] reported WE43 alloys to show no toxicity for osteoblasts, gingival fibroblasts and sarcoma cells. These results agreed with those reported in ref. [424], where MC3T3-E1 cells were reported to have an elongated and spread morphology on Mg–Y–Ca–Zr alloy. In addition, *in vivo* studies reported Mg–Y–Ca–Zr

TABLE 3.43

Summary of the corrosion rates for Mg–RE alloys.*

Alloy	Additional elements	Corrosive environment	Immersion time	Corrosion rate (mm/year)	Procedure	Ref.
Mg–1Y–0.6Ca–0.4Zr	0.6 wt.% Ca + 0.4 wt.% Zr	DMEM + 10% FBS, 37°C	1 week	0.49	WL	[424]
Mg–1Y–0.6Ca–0.4Zr (T4)	0.6 wt.% Ca + 0.4 wt.% Zr	DMEM + 10% FBS, 37°C	1 week	0.56	WL	[424]
Mg–4Y–0.6Ca–0.4Zr	0.6 wt.% Ca + 0.4 wt.% Zr	DMEM + 10% FBS, 37°C	1 week	0.57	WL	[424]
Mg–4Y–0.6Ca–0.4Zr (T4)	0.6 wt.% Ca + 0.4 wt.% Zr	DMEM + 10% FBS, 37°C	1 week	0.58	WL	[424]
Mg–1Y–0.6Ca–0.4Zr	0.6 wt.% Ca + 0.4 wt.% Zr	DMEM + 10% FBS, 37°C	2 weeks	0.69	WL	[424]
Mg–1Y–0.6Ca–0.4Zr (T4)	0.6 wt.% Ca + 0.4 wt.% Zr	DMEM + 10% FBS, 37°C	2 weeks	0.79	WL	[424]
Mg–4Y–0.6Ca–0.4Zr	0.6 wt.% Ca + 0.4 wt.% Zr	DMEM + 10% FBS, 37°C	2 weeks	0.41	WL	[424]
Mg–4Y–0.6Ca–0.4Zr (T4)	0.6 wt.% Ca + 0.4 wt.% Zr	DMEM + 10% FBS, 37°C	2 weeks	0.61	WL	[424]
Mg–1Y–0.6Ca–0.4Zr	0.6 wt.% Ca + 0.4 wt.% Zr	DMEM + 10% FBS, 37°C	3 weeks	0.72	WL	[424]
Mg–1Y–0.6Ca–0.4Zr (T4)	0.6 wt.% Ca + 0.4 wt.% Zr	DMEM + 10% FBS, 37°C	3 weeks	0.64	WL	[424]
Mg–4Y–0.6Ca–0.4Zr	0.6 wt.% Ca + 0.4 wt.% Zr	DMEM + 10% FBS, 37°C	3 weeks	0.46	WL	[424]
Mg–4Y–0.6Ca–0.4Zr (T4)	0.6 wt.% Ca + 0.4 wt.% Zr	DMEM + 10% FBS, 37°C	3 weeks	0.55	WL	[424]
Mg–1Y–0.6Ca–0.4Zr	0.6 wt.% Ca + 0.4 wt.% Zr	*In vivo*, mice	40 days	0.31	WL	[424]
Mg–4Y–0.6Ca–0.4Zr	0.6 wt.% Ca + 0.4 wt.% Zr	*In vivo*, mice	40 days	0.18	WL	[424]
Mg–1Y–0.6Ca–0.4Zr	0.6 wt.% Ca + 0.4 wt.% Zr	*In vivo*, mice	70 days	0.73	WL	[424]
Mg–4Y–0.6Ca–0.4Zr	0.6 wt.% Ca + 0.4 wt.% Zr	*In vivo*, mice	70 days	0.068	WL	[424]
Mg–1Gd	—	9 g/l, 37°C	—	24.7	WL	[425]
Mg–3Gd	—	9 g/l, 37°C	—	0.29	WL	[425]

(Continued)

TABLE 3.43 (Cont.)

Alloy	Additional elements	Corrosive environment	Immersion time	Corrosion rate (mm/year)	Procedure	Ref.
Mg–5Gd	—	9 g/l, 37°C	—	0.38	WL	[425]
Mg–5Gd (T4)	—	9 g/l, 37°C	—	23.5	WL	[425]
Mg–1Nd	—	9 g/l, 37°C	—	1.7	WL	[425]
Mg–4Nd	—	9 g/l, 37°C	—	1.0	WL	[425]
Mg–4Nd (T4)	—	9 g/l, 37°C	—	0.67	WL	[425]
Mg–9Nd	—	9 g/l, 37°C	—	29.0	WL	[425]
Mg–1Gd–1Y	—	9 g/l, 37°C	—	0.93	WL	[425]
Mg–3Gd–1Y	—	9 g/l, 37°C	—	3.2	WL	[425]
Mg–5Gd–1Y	—	9 g/l, 37°C	—	6.6	WL	[425]

* All the Mg–RE alloys are cast, unless otherwise specified. Additional information can be found on www.routledge.com/9780367429454.

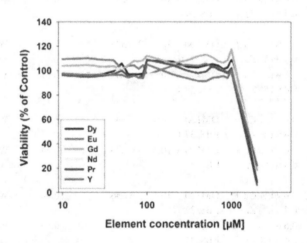

FIGURE 3.17 Viability of MG63 subjected to different RE element. Reprinted with permission from Elsevier [363].

alloy to not present an acceptable host response because of a dense connective tissue adjacent to the Mg alloy implants. Moreover, their *in vitro* cytocompatibility and *in vivo* host response were found to be better compared to pure Mg and AZ31. Finally, Castellani et al. [428] reported a modified WE43 alloy to have a bone-implant interfacial strength higher than titanium controls. However, Ye et al. reported WE43 alloy to have a viability for L-929 cells of about 30% [429–431]. This low biocompatibility agreed with the thrombosis reported in the case of WE43. The biocompatibility of Mg–RE alloys was reported to be improved after T4 and T6 treatments and after grain refinement by SPD techniques [432,433], and further results can be found in Table 3.44.

TABLE 3.44

Summary of the biocompatibility of Mg–RE alloys.*

			Cell viability				
Alloy	Additional elements (wt.%)	Test environment	Cell type, procedure	Time of culture	Result (%)	Hemolysis (%)	Ref.
WE43	—	DMEM, 37°C	Osteoblast, IC	1 day	108.7	—	[427]
WE43	—	DMEM, 37°C	Fibroblast, IC	1 day	105.6	—	[427]
WE43	—	DMEM, 37°C	SaOS-2, IC	1 day	98.9	—	[427]
Mg–1Y–0.6Ca–0.4Zr	0.6 wt.% Ca + 0.4 wt.% Zr	α-MEM, 37°C	MC3T3-E1, IC	1 day	142.5	—	[424]
Mg–1Y–0.6Ca–0.4Zr (T4)	0.6 wt.% Ca + 0.4 wt.% Zr	α-MEM, 37°C	MC3T3-E1, IC	1 day	150.4	—	[424]
Mg–4Y–0.6Ca–0.4Zr	0.6 wt.% Ca + 0.4 wt.% Zr	α-MEM, 37°C	MC3T3-E1, IC	1 day	133.5	—	[424]
Mg–4Y–0.6Ca–0.4Zr (T4)	0.6 wt.% Ca + 0.4 wt.% Zr	α-MEM, 37°C	MC3T3-E1, IC	1 day	149.2	—	[424]
Mg–1Y–0.6Ca–0.4Zr	0.6 wt.% Ca + 0.4 wt.% Zr	α-MEM, 37°C	MC3T3-E1, IC	3 days	110.4	—	[424]
Mg–1Y–0.6Ca–0.4Zr (T4)	0.6 wt.% Ca + 0.4 wt.% Zr	α-MEM, 37°C	MC3T3-E1, IC	3 days	98.3	—	[424]
Mg–4Y–0.6Ca–0.4Zr	0.6 wt.% Ca + 0.4 wt.% Zr	α-MEM, 37°C	MC3T3-E1, IC	3 days	93.8	—	[424]
Mg–4Y–0.6Ca–0.4Zr (T4)	0.6 wt.% Ca + 0.4 wt.% Zr	α-MEM, 37°C	MC3T3-E1, IC	3 days	115.5	—	[424]
WE43	—	DMEM, 37°C	L-929, IC	1 day	25.1	—	[429–431]
WE43	—	DMEM, 37°C	L-929, IC	2 days	34.7	—	[429–431]
WE43	—	DMEM, 37°C	L-929, IC	3 days	34.8	—	[429–431]
WE43 (extruded)	—	Diluted blood, 37°C	—	—	—	9.3	[429–431]

* All the Mg–RE alloys are cast, unless otherwise specified. Additional information can be found on www.routledge.com/9780367429454.

REFERENCES

[1] Hofstetter, J., Martinelli, E., Weinberg, A.M., Becker, M., Mingler, B., Uggowitzer, P.J., Löffler, J.F. (2015). Assessing the degradation performance of ultrahigh-purity magnesium in vitro and *in vivo*, *Corros. Sci.*, 91, pp. 29–36, Doi: 10.1016/j.corsci.2014.09.008.

[2] Jeong, Y.S., Kim, W.J. (2014). Enhancement of mechanical properties and corrosion resistance of Mg–Ca alloys through microstructural refinement by indirect extrusion, *Corros. Sci.*, 82, pp. 392–403, Doi: 10.1016/J.CORSCI.2014.01.041.

[3] Han, H.-S., Lee, S.H., Kim, W.-J., Jeon, H., Seok, H.-K., Ahn, J.-P., Kim, Y.-C. (2015). Reduction of initial corrosion rate and improvement of cell adhesion through surface modification of biodegradable Mg alloy, *Met. Mater. Int.*, 21(1), pp. 194–201, Doi: 10.1007/s12540-015-1024-6.

[4] Gu, X., Zhou, W., Zheng, Y., Dong, L., Xi, Y., Chai, D. (2010). Microstructure, mechanical property, bio-corrosion and cytotoxicity evaluations of Mg/HA composites, *Mater. Sci. Eng. C*, 30(6), pp. 827–32, Doi: 10.1016/J.MSEC. 2010.03.016.

[5] Seong, J.W., Kim, W.J. (2015). Development of biodegradable Mg–Ca alloy sheets with enhanced strength and corrosion properties through the refinement and uniform dispersion of the Mg2Ca phase by high-ratio differential speed rolling, *Acta Biomater.*, 11, pp. 531–42, Doi: 10.1016/J.ACTBIO.2014.09.029.

[6] Pachla, W., Mazur, A., Skiba, J., Kulczyk, M., Przybysz, S. (2012). Development of high-strength pure magnesium and wrought magnesium alloys AZ31, AZ61, and AZ91 processed by hydrostatic extrusion with back pressure, *Int. J. Mater. Res.*, 103(5), pp. 580–9, Doi: 10.3139/146.110721.

[7] Li, Z., Huang, N., Zhao, J., Zhou, S.J. (2013). Microstructure, mechanical and degradation properties of equal channel angular pressed pure magnesium for biomedical application, *Mater. Sci. Technol.*, 29(2), pp. 140–7, Doi: 10.1179/1743284712Y.0000000148.

[8] Gan, W.M., Zheng, M.Y., Chang, H., Wang, X.J., Qiao, X.G., Wu, K., Schwebke, B., Brokmeier, H.-G. (2009). Microstructure and tensile property of the ECAPed pure magnesium, *J. Alloys Compd.*, 470(1–2), pp. 256–62, Doi: 10.1016/J. JALLCOM.2008.02.030.

[9] Xiong, G., Nie, Y., Ji, D., Li, J., Li, C., Li, W., Zhu, Y., Luo, H., Wan, Y. (2016). Characterization of biomedical hydroxyapatite/magnesium composites prepared by powder metallurgy assisted with microwave sintering, *Curr. Appl. Phys.*, 16(8), pp. 830–6, Doi: 10.1016/J.CAP.2016.05.004.

[10] Campo, R.D., Savoini, B., Muñoz, A., Monge, M.A., Garcés, G. (2014). Mechanical properties and corrosion behavior of Mg–HAP composites, *J. Mech. Behav. Biomed. Mater.*, 39, pp. 238–46, Doi: 10.1016/j.jmbbm.2014.07.014.

[11] Saha, P., Roy, M., Datta, M.K., Lee, B., Kumta, P.N. (2015). Effects of grain refinement on the biocorrosion and in vitro bioactivity of magnesium, *Mater. Sci. Eng. C*, 57, pp. 294–303, Doi: 10.1016/j.msec.2015.07.033.

[12] Bowen, P.K., Drelich, J., Goldman, J. (2013). A new in vitro–in vivo correlation for bioabsorbable magnesium stents from mechanical behavior, *Mater. Sci. Eng. C*, 33, pp. 5064–70, Doi: 10.1016/j.msec.2013.08.042.

[13] Yang, C., Yuan, G., Zhang, J., Tang, Z., Zhang, X., Dai, K. (2010). Effects of magnesium alloys extracts on adult human bone marrow-derived stromal cell viability and osteogenic differentiation, *Biomed. Mater.*, 5(4), p. 045005, Doi: 10.1088/1748-6041/5/4/045005.

[14] Witte, F., Kaese, V., Haferkamp, H., Switzer, E., Meyer-Lindenberg, A., Wirth, C. J., Windhagen, H. (2005). In vivo corrosion of four magnesium alloys and the

associated bone response, *Biomaterials*, 26(17), pp. 3557–63, Doi: 10.1016/j.
biomaterials.2004.09.049.

[15] Song, G. (2007). Control of biodegradation of biocompatable magnesium alloys,
Corros. Sci., 49(4), pp. 1696–701, Doi: 10.1016/J.CORSCI.2007.01.001.

[16] Liu, M., Uggowitzer, P.J., Nagasekhar, A.V., Schmutz, P., Easton, M.,
Song, G.-L., Atrens, A. (2009). Calculated phase diagrams and the corrosion
of die-cast Mg–Al alloys, *Corros. Sci.*, 51(3), pp. 602–19, Doi: 10.1016/J.
CORSCI.2008.12.015.

[17] Song, G.-L., Atrens, A. (1999). Corrosion mechanisms of magnesium alloys, *Adv.
Eng. Mater.*, 1(1), pp. 11–33, Doi: 10.1002/(SICI)1527-2648(199909)1.

[18] Reichek, K.N., Clark, K.J., Hillis, J.E. (1985). Controlling the salt water corrosion
performance of magnesium AZ91 alloy, *SAE Trans.*, 94, pp. 318–29, Doi:
10.2307/44721571.

[19] Lunder, O., Aune, T.K., Nisancioglu, K. (1987). Effect of Mn additions on the corrosion behavior of mould-cast magnesium ASTM AZ91, *Corrosion*, 43(5), pp.
291–5, Doi: 10.5006/1.3583151.

[20] Lee, J.-Y., Han, G., Kim, Y.-C., Byun, J.-Y., Jang, J., Seok, H.-K., Yang, S.-J.
(2009). Effects of impurities on the biodegradation behavior of pure magnesium,
Met. Mater. Int., 15(6), pp. 955–61, Doi: 10.1007/s12540-009-0955-1.

[21] Zainal Abidin, N.I., Atrens, A.D., Martin, D., Atrens, A. (2011). Corrosion of
high purity Mg, Mg2Zn0.2Mn, ZE41 and AZ91 in Hank's solution at 37 °C,
Corros. Sci., 53(11), pp. 3542–56, Doi: 10.1016/J.CORSCI.2011.06.030.

[22] Li, X., Liu, X., Wu, S., Yeung, K.W.K., Zheng, Y., Chu, P.K. (2016). Design of
magnesium alloys with controllable degradation for biomedical implants: From
bulk to surface, *Acta Biomater.*, 45, pp. 2–30, Doi: 10.1016/J.ACTBIO.2016.09.005.

[23] Zhang, S., Zheng, Y., Zhang, L., Bi, Y., Li, J., Liu, J., Yu, Y., Guo, H., Li, Y.
(2016). In vitro and in vivo corrosion and histocompatibility of pure Mg and a
Mg-6Zn alloy as urinary implants in rat model, *Mater. Sci. Eng. C*, 68, pp. 414–
22, Doi: 10.1016/J.MSEC.2016.06.017.

[24] Gu, X., Zheng, Y., Cheng, Y., Zhong, S., Xi, T. (2009). In vitro corrosion and biocompatibility of binary magnesium alloys, *Biomaterials*, 30(4), pp. 484–98, Doi:
10.1016/J.BIOMATERIALS.2008.10.021.

[25] Walker, J., Shadanbaz, S., Kirkland, N.T., Stace, E., Woodfield, T., Staiger, M.P.,
Dias, G.J. (2012). Magnesium alloys: Predicting in vivo corrosion with in vitro
immersion testing, *J. Biomed. Mater. Res. Part B Appl. Biomater.*, 100, Doi:
10.1002/jbm.b.32680.

[26] Shadanbaz, S., Walker, J., Staiger, M.P., Dias, G.J., Pietak, A. (2013). Growth of
calcium phosphates on magnesium substrates for corrosion control in biomedical
applications via immersion techniques, *J. Biomed. Mater. Res. Part B Appl. Biomater.*, 101B(1), pp. 162–72, Doi: 10.1002/jbm.b.32830.

[27] Zhou, Y.-L., Li, Y., Luo, D.-M., Ding, Y., Hodgson, P. (2015). Microstructures,
mechanical and corrosion properties and biocompatibility of as extruded Mg–
Mn–Zn–Nd alloys for biomedical applications, *Mater. Sci. Eng. C*, 49, pp. 93–
100, Doi: 10.1016/j.msec.2014.12.057.

[28] Nguyen, T.L., Blanquet, A., Staiger, M.P., Dias, G.J., Woodfield, T.B.F. (2012).
On the role of surface roughness in the corrosion of pure magnesium in vitro,
J. Biomed. Mater. Res. Part B Appl. Biomater., 100B(5), pp. 1310–8, Doi: 10.1002/
jbm.b.32697.

[29] Murr, E.L., Gaytan, M.S., Martinez, E., Medina, F.R., Wicker, R.B. (2012). Fab-
ricating functional Ti-alloy biomedical implants by additive manufacturing using
electron beam melting, *J. Biotechnol. Biomater.*, 2(3), Doi: 10.4172/2155-
952X.1000131.

[30] Obaton, A.-F., Fain, J., Djemaï, M., Meinel, D., Léonard, F., Mahé, E., Lécuelle, B., Fouchet, J.-J., Bruno, G. (2017). In vivo XCT bone characterization of lattice structured implants fabricated by additive manufacturing, *Heliyon*, 3(8), Doi: 10.1016/J.HELIYON.2017.E00374.

[31] Smeets, R., Stadlinger, B., Schwarz, F., Beck-Broichsitter, B., Jung, O., Precht, C., Kloss, F., Gröbe, A., Heiland, M., Ebker, T., Ebker, T. (2016). Impact of dental implant surface modifications on osseointegration, *Biomed. Res. Int.*, 2016, pp. 1–16, Doi: 10.1155/2016/6285620.

[32] Albrektsson, T., Wennerberg, A. (n.d.). Oral implant surfaces: Part 1–Review focusing on topographic and chemical properties of different surfaces and in vivo responses to them, *Int. J. Prosthodont.*, 17(5), pp. 536–43.

[33] Nguyen, T., Waterman, J., Staiger, M., Woodfield, T. (2012). Controlling *in vitro* corrosion rate of pure Mg with rough surface texture via biomimetic coating systems, *Corros. Eng. Sci. Technol.*, 47(5), pp. 358–64, Doi: 10.1179/1743278212Y. 0000000023.

[34] Li, R.W., Kirkland, N.T., Truong, J., Wang, J., Smith, P.N., Birbilis, N., Nisbet, D. R. (2014). The influence of biodegradable magnesium alloys on the osteogenic differentiation of human mesenchymal stem cells, *J. Biomed. Mater. Res. Part A*, 102 (12), pp. n/a–n/a, Doi: 10.1002/jbm.a.35111.

[35] Pietak, A., Mahoney, P., Dias, G.J., Staiger, M.P. (2008). Bone-like matrix formation on magnesium and magnesium alloys, *J. Mater. Sci. Mater. Med.*, 19(1), pp. 407–15, Doi: 10.1007/s10856-007-3172-9.

[36] Schumacher, S., Roth, I., Stahl, J., Bäumer, W., Kietzmann, M. (2014). Biodegradation of metallic magnesium elicits an inflammatory response in primary nasal epithelial cells, *Acta Biomater.*, 10(2), pp. 996–1004, Doi: 10.1016/J.ACTBIO. 2013.10.030.

[37] Kainer, K.U., Bala Srinivasan, P., Blawert, C., Dietzel, W. (2010). Corrosion of magnesium and its alloys. In *Shreir's corrosion*, Ed. J.A.R. Tony, Oxford, Elsevier, pp. 2011–41.

[38] Song, G., Atrens, A., Wu, X., Zhang, B. (1998). Corrosion behaviour of AZ21, AZ501 and AZ91 in sodium chloride, *Corros. Sci.*, 40(10), pp. 1769–91, Doi: 10.1016/S0010-938X(98)00078-X.

[39] Gusieva, K., Davies, C.H.J., Scully, J.R., Birbilis, N. (n.d.). Corrosion of magnesium alloys: The role of alloying, Doi: 10.1179/1743280414Y.0000000046.

[40] Lunder, O., Lein, J.E., Aune, T.K., Nisancioglu, K. (1989). The role of Mg $_{17}$ Al $_{12}$ phase in the corrosion of Mg alloy AZ91, *Corrosion*, 45(9), pp. 741–8, Doi: 10.5006/1.3585029.

[41] Hermann, F., Sommer, F., Jones, H., Edyvean, R.G.J. (1989). Corrosion inhibition in magnesium-aluminium-based alloys induced by rapid solidification processing, *J. Mater. Sci.*, 24(7), pp. 2369–79, Doi: 10.1007/BF01174498.

[42] Winzer, N., Atrens, A., Song, G., Ghali, E., Dietzel, W., Kainer, K.U., Hort, N., Blawert, C. (2005). A critical review of the Stress Corrosion Cracking (SCC) of magnesium alloys, *Adv. Eng. Mater.*, 7(8), pp. 659–93, Doi: 10.1002/ adem.200500071.

[43] Domingo, J.L. (1995). Reproductive and developmental toxicity of aluminum: A review, *Neurotoxicol. Teratol.*, 17(4), pp. 515–21, Doi: 10.1016/0892-0362(95) 00002-9.

[44] Venugopal, B., Luckey, T.D. (1978). *Metal toxicity in mammals. Volume 2. Chemical toxicity of metals and metalloids*, New York, Plenum Press.

[45] Flaten, T.P. (2001). Aluminium as a risk factor in Alzheimer's disease, with emphasis on drinking water, *Brain Res. Bull.*, 55(2), pp. 187–96.

[46] El-Rahman, S.S.A. (2003). Neuropathology of aluminum toxicity in rats (glutamate and GABA impairment)., *Pharmacol. Res.*, 47(3), pp. 189–94.

[47] Lü, Y., Wang, Q., Zeng, X., Ding, W., Zhai, C., Zhu, Y. (2000). Effects of rare earths on the microstructure, properties and fracture behavior of Mg–Al alloys, *Mater. Sci. Eng. A*, 278(1–2), pp. 66–76, Doi: 10.1016/S0921-5093(99)00604-8.

[48] Zhou, H., Zeng, X., Liu, L., Zhang, Y., Zhu, Y., Ding, W. (2004). Effect of cerium on microstructures and mechanical properties of AZ61 wrought magnesium alloy, *J. Mater. Sci.*, 39(23), pp. 7061–6, Doi: 10.1023/B:JMSC.0000047551.04037.fe.

[49] Akyuz, B. (2013). Influence of Al content on machinability of AZ series Mg alloys, *Trans. Nonferrous Met. Soc. China*, 23(8), pp. 2243–9, Doi: 10.1016/S1003-6326(13)62724-7.

[50] Wang, Y., Zhou, J., Wang, J., Luo, T., Yang, Y. (2011). Effect of Bi addition on microstructures and mechanical properties of AZ80 magnesium alloy, *Trans. Nonferrous Met. Soc. China*, 21(4), pp. 711–6, Doi: 10.1016/S1003-6326(11)60770-X.

[51] Candan, S., Unal, M., Koc, E., Turen, Y., Candan, E. (2011). Effects of titanium addition on mechanical and corrosion behaviours of AZ91 magnesium alloy, *J. Alloys Compd.*, 509(5), pp. 1958–63, Doi: 10.1016/J.JALLCOM.2010.10.100.

[52] Zberg, B., Uggowitzer, P.J., Löffler, J.F. (2009). MgZnCa glasses without clinically observable hydrogen evolution for biodegradable implants, *Nat. Mater.*, 8, Doi: 10.1038/NMAT2542.

[53] Zberg, B., Arata, E.R., Uggowitzer, P.J., Löffler, J.F. (2009). Tensile properties of glassy MgZnCa wires and reliability analysis using Weibull statistics, *Acta Mater.*, 57(11), pp. 3223–31, Doi: 10.1016/J.ACTAMAT.2009.03.028.

[54] Liu, L., Yuan, F., Zhao, M., Gao, C., Feng, P., Yang, Y., Yang, S., Shuai, C. (2017). Rare earth element yttrium modified Mg-Al-Zn alloy: Microstructure, degradation properties and hardness, *Materials (Basel)*, 10(5), p. 477, Doi: 10.3390/ma10050477.

[55] Luo, T.J., Yang, Y.S., Li, Y.J., Dong, X.G. (2009). Influence of rare earth Y on the corrosion behavior of as-cast AZ91 alloy, *Electrochim. Acta*, 54(26), pp. 6433–7, Doi: 10.1016/J.ELECTACTA.2009.06.023.

[56] Wu, D., Yan, S., Wang, Z., Zhang, Z., Miao, R., Zhang, X., Chen, D. (2014). Effect of samarium on microstructure and corrosion resistance of aged as-cast AZ92 magnesium alloy, *J. Rare Earths*, 32(7), pp. 663–71, Doi: 10.1016/S1002-0721(14)60123-X.

[57] Wang, X., Cai, S., Xu, G., Ye, X., Ren, M., Huang, K. (2013). Surface characteristics and corrosion resistance of sol–gel derived CaO–P2O5–SrO–Na2O bioglass–Ceramic coated Mg alloy by different heat-treatment temperatures, *J. Sol-Gel Sci. Technol.*, 67(3), pp. 629–38, Doi: 10.1007/s10971-013-3122-6.

[58] Wu, G., Xu, R., Feng, K., Wu, S., Wu, Z., Sun, G., Zheng, G., Li, G., Chu, P.K. (2012). Retardation of surface corrosion of biodegradable magnesium-based materials by aluminum ion implantation, *Appl. Surf. Sci.*, 258(19), pp. 7651–7, Doi: 10.1016/J.APSUSC.2012.04.112.

[59] Qiu, X., Wan, P., Tan, L., Fan, X., Yang, K. (2014). Preliminary research on a novel bioactive silicon doped calcium phosphate coating on AZ31 magnesium alloy via electrodeposition, *Mater. Sci. Eng. C*, 36, pp. 65–76, Doi: 10.1016/J.MSEC.2013.11.041.

[60] Gu, X.N., Zheng, Y.F., Chen, L.J. (2009). Influence of artificial biological fluid composition on the biocorrosion of potential orthopedic Mg–Ca, AZ31, AZ91 alloys, *Biomed. Mater.*, 4(6), p. 065011, Doi: 10.1088/1748-6041/4/6/065011.

[61] Ratna Sunil, B., Sampath Kumar, T.S., Chakkingal, U., Nandakumar, V., Doble, M., Devi Prasad, V., Raghunath, M. (2016). In vitro and in vivo studies of biodegradable fine grained AZ31 magnesium alloy produced by equal channel

angular pressing, *Mater. Sci. Eng. C*, 59, pp. 356–67, Doi: 10.1016/J. MSEC.2015.10.028.

[62] Ostrowski, N., Lee, B., Enick, N., Carlson, B., Kunjukunju, S., Roy, A., Kumta, P. N. (2013). Corrosion protection and improved cytocompatibility of biodegradable polymeric layer-by-layer coatings on AZ31 magnesium alloys, *Acta Biomater.*, 9 (10), pp. 8704–13, Doi: 10.1016/J.ACTBIO.2013.05.010.

[63] Wen, C., Guan, S., Peng, L., Ren, C., Wang, X., Hu, Z. (2009). Characterization and degradation behavior of AZ31 alloy surface modified by bone-like hydroxy-apatite for implant applications, *Appl. Surf. Sci.*, 255(13–14), pp. 6433–8, Doi: 10.1016/J.APSUSC.2008.09.078.

[64] Hosaka, T., Yoshihara, S., Amanina, I., MacDonald, B.J. (2017). Influence of grain refinement and residual stress on corrosion behavior of AZ31 magnesium alloy processed by ECAP in RPMI-1640 medium, *Procedia Eng.*, 184, pp. 432–41, Doi: 10.1016/J.PROENG.2017.04.114.

[65] Zhu, Y., Wu, G., Zhang, Y.-H., Zhao, Q. (2011). Growth and characterization of Mg(OH)2 film on magnesium alloy AZ31, *Appl. Surf. Sci.*, 257(14), pp. 6129–37, Doi: 10.1016/J.APSUSC.2011.02.017.

[66] Bukovinová, L., Hadzima, B. (2012). Electrochemical characteristics of magnesium alloy AZ31 in Hank's solution, *Corros. Eng. Sci. Technol.*, 47(5), pp. 352–7, Doi: 10.1179/1743278212Y.0000000033.

[67] Wang, H., Estrin, Y., Zúberová, Z. (2008). Bio-corrosion of a magnesium alloy with different processing histories, *Mater. Lett.*, 62(16), pp. 2476–9, Doi: 10.1016/ J.MATLET.2007.12.052.

[68] Gu, X.N., Li, N., Zheng, Y.F., Kang, F., Wang, J.T., Ruan, L. (2011). In vitro study on equal channel angular pressing AZ31 magnesium alloy with and without back pressure, *Mater. Sci. Eng. B*, 176(20), pp. 1802–6, Doi: 10.1016/J. MSEB.2011.04.003.

[69] Bagherifard, S., Hickey, D.J., Fintová, S., Pastorek, F., Fernandez-Pariente, I., Bandini, M., Webster, T.J., Guagliano, M. (2018). Effects of nanofeatures induced by severe shot peening (SSP) on mechanical, corrosion and cytocompatibility properties of magnesium alloy AZ31, *Acta Biomater.*, 66, pp. 93–108, Doi: 10.1016/J.ACTBIO.2017.11.032.

[70] Xin, R., Luo, Y., Zuo, A., Gao, J., Liu, Q. (2012). Texture effect on corrosion behavior of AZ31 Mg alloy in simulated physiological environment, *Mater. Lett.*, 72, pp. 1–4, Doi: 10.1016/J.MATLET.2011.11.032.

[71] Bertolini, R., Bruschi, S., Ghiotti, A., Pezzato, L., Dabalà, M. (2017). The effect of cooling strategies and machining feed rate on the corrosion behavior and wett-ability of AZ31 alloy for biomedical applications, *Procedia CIRP*, 65, pp. 7–12, Doi: 10.1016/J.PROCIR.2017.03.168.

[72] Adekanmbi, I., Mosher, C.Z., Lu, H.H., Riehle, M., Kubba, H., Tanner, K.E. (2017). Mechanical behaviour of biodegradable AZ31 magnesium alloy after long term in vitro degradation, *Mater. Sci. Eng. C*, 77, pp. 1135–44, Doi: 10.1016/j. msec.2017.03.216.

[73] Wang, S.-H., Yang, C.-W., Lee, T.-M. (2016). Evaluation of microstructural features and *in vitro* biocompatibility of hydrothermally coated fluorohydroxyapatite on AZ80 Mg alloy, *Ind. Eng. Chem. Res.*, 55(18), pp. 5207–15, Doi: 10.1021/acs. iecr.5b04583.

[74] Gong, H., Kontsos, A., Kim, Y., Lelkes, P.I., Zhang, Q., Yao, D., Hazeli, K., Zhou, J.G. (2012). Micro characterization of Mg and Mg alloy for biodegradable orthopedic implants application, ASME 2012 International Manufacturing Science and Engineering Conference, ASME, p. 891.

[75] Razavi, M., Fathi, M., Savabi, O., Vashaee, D., Tayebi, L. (2015). *In vivo* study of nanostructured akermanite/PEO coating on biodegradable magnesium alloy for biomedical applications, *J. Biomed. Mater. Res. Part A*, 103(5), pp. 1798–808, Doi: 10.1002/jbm.a.35324.

[76] Razavi, M., Fathi, M., Savabi, O., Vashaee, D., Tayebi, L. (2014). In vitro study of nanostructured diopside coating on Mg alloy orthopedic implants, *Mater. Sci. Eng. C*, 41, pp. 168–77, Doi: 10.1016/j.msec.2014.04.039.

[77] Kim, J., Mousa, H.M., Park, C.H., Kim, C.S. (2017). Enhanced corrosion resistance and biocompatibility of AZ31 Mg alloy using PCL/ZnO NPs via electrospinning, *Appl. Surf. Sci.*, 396, pp. 249–58, Doi: 10.1016/J.APSUSC.2016.10.092.

[78] Hiromoto, S., Yamazaki, T. (2017). Micromorphological effect of calcium phosphate coating on compatibility of magnesium alloy with osteoblast, *Sci. Technol. Adv. Mater.*, 18(1), pp. 96–109, Doi: 10.1080/14686996.2016.1266238.

[79] Monfoulet, L.-E., Becquart, P., Marchat, D., Vandamme, K., Bourguignon, M., Pacard, E., Viateau, V., Petite, H., Logeart-Avramoglou, D. (2014). The pH in the microenvironment of human mesenchymal stem cells is a critical factor for optimal osteogenesis in tissue-engineered constructs, *Tissue Eng. Part A*, 20(13–14), pp. 1827–40, Doi: 10.1089/ten.tea.2013.0500.

[80] Lang, F., Föller, M., Lang, K., Lang, P., Ritter, M., Vereninov, A., Szabo, I., Huber, S.M., Gulbins, E. (2007). Cell volume regulatory ion channels in cell proliferation and cell death, *Methods Enzymol.*, 428, pp. 209–25, Doi: 10.1016/S0076-6879(07)28011-5.

[81] Wong, H.M., Wu, S., Chu, P.K., Cheng, S.H., Luk, K.D.K., Cheung, K.M.C., Yeung, K.W.K. (2013). Low-modulus Mg/PCL hybrid bone substitute for osteoporotic fracture fixation, *Biomaterials*, 34(29), pp. 7016–32, Doi: 10.1016/J.BIOMATERIALS.2013.05.062.

[82] Luo, A.A., Zhang, C., Sachdev, A.K. (2012). Effect of eutectic temperature on the extrudability of magnesium–aluminum alloys, *Scr. Mater.*, 66(7), pp. 491–4, Doi: 10.1016/J.SCRIPTAMAT.2011.12.025.

[83] Robinson, H.A., George, P.F. (1954). Effect of alloying and impurity elements in magnesium alloy cast anodes, *Corrosion*, 10(6), pp. 182–8, Doi: 10.5006/0010-9312-10.6.182.

[84] Culotta, V.C., Yang, M., Hall, M.D. (2005). Manganese transport and trafficking: Lessons learned from Saccharomyces cerevisiae, *Eukaryot. Cell*, 4(7), pp. 1159–65, Doi: 10.1128/EC.4.7.1159-1165.2005.

[85] Huang, W., Yan, H. (2011). Effect of cerium addition on microstructure and mechanical properties of die-cast AM60 alloy. Proceedings of 2011 International Conference on Electronic & Mechanical Engineering and Information Technology, IEEE, pp. 2609–11.

[86] Braszczyńska-Malik, K.N., Grzybowska, A. (2016). Influence of phase composition on microstructure and properties of Mg-5Al-0.4Mn-xRE (x = 0, 3 and 5 wt.%) alloys, *Mater. Charact.*, 115, pp. 14–22, Doi: 10.1016/J.MATCHAR.2016.03.014.

[87] Kang, S.B., Cho, J., Chang, L., Wang, Y. (2011). Influence of twin roll casting and differential speed rolling on microstructure and tensile properties in magnesium alloy sheets, *Procedia Eng.*, 10, pp. 1190–5, Doi: 10.1016/J.PROENG.2011.04.198.

[88] Mert, F., Özdemir, A., Kainer, K.U., Hort, N. (2013). Influence of Ce addition on microstructure and mechanical properties of high pressure die cast AM50 magnesium alloy, *Trans. Nonferrous Met. Soc. China*, 23(1), pp. 66–72, Doi: 10.1016/S1003-6326(13)62430-9.

[89] Wang, M., Zhou, H., Wang, L. (2007). Effect of yttrium and cerium addition on microstructure and mechanical properties of AM50 magnesium alloy, *J. Rare Earths*, 25(2), pp. 233–7, Doi: 10.1016/S1002-0721(07)60079-9.

[90] Cui, X., Liu, H., Meng, J., Zhang, D. (2010). Microstructure and mechanical properties of die-cast AZ91D magnesium alloy by Pr additions, *Trans. Nonferrous Met. Soc. China*, 20, pp. s435–8, Doi: 10.1016/S1003-6326(10)60513-4.

[91] Liu, W., Cao, F., Chang, L., Zhang, Z., Zhang, J. (2009). Effect of rare earth element Ce and La on corrosion behavior of AM60 magnesium alloy, *Corros. Sci.*, 51 (6), pp. 1334–43, Doi: 10.1016/j.corsci.2009.03.018.

[92] Braszczyńska-Malik, K.N. (2017). Effect of high-pressure die casting on structure and properties of Mg-5Al-0.4Mn-xRE (x = 1, 3 and 5 wt%) experimental alloys, *J. Alloys Compd.*, 694, pp. 841–7, Doi: 10.1016/J.JALLCOM.2016.10.033.

[93] Su, G., Zhang, L., Cheng, L., Liu, Y., Cao, Z. (2010). Microstructure and mechanical properties of Mg-6Al-0.3Mn-xY alloys prepared by casting and hot rolling, *Trans. Nonferrous Met. Soc. China*, 20(3), pp. 383–9, Doi: 10.1016/S1003-6326(09)60150-3.

[94] Ding, H.-L., Zhang, Y.-W., Kamado, S. (2014). Effect of finish-rolling conditions on mechanical properties and texture characteristics of AM50 alloy sheet, *Trans. Nonferrous Met. Soc. China*, 24, pp. 2761–6, Doi: 10.1016/S1003-6326(14)63407-5.

[95] Sahu, P.K., Pal, S. (2018). Effect of FSW parameters on microstructure and mechanical properties of AM20 welds, *Mater. Manuf. Process.*, 33(3), pp. 288–98, Doi: 10.1080/10426914.2017.1279295.

[96] Yu, Z., Hu, M., Tang, A., Wu, M., He, J., Gao, Z., Wang, F., Li, C., Chen, B., Liu, J. (2017). Effect of aluminium on the microstructure and mechanical properties of as-cast magnesium–manganese alloys, *Mater. Sci. Technol.*, 33(17), pp. 2086–96, Doi: 10.1080/02670836.2017.1345824.

[97] Anawati, A., Asoh, H., Ono, S. (2016). Effects of alloying element Ca on the corrosion behavior and bioactivity of anodic films formed on AM60 Mg alloys, *Materials* (Basel)., 10(1), p. 11, Doi: 10.3390/ma10010011.

[98] Abdal-Hay, A., Hasan, A., Kim, Y.K., Lee, M.-H., Hamdy, A.S., Khalil, K.A. (2016). Biocorrosion behavior of biodegradable nanocomposite fibers coated layer-by-layer on AM50 magnesium implant, *Mater. Sci. Eng. C*, 58, pp. 1232–41, Doi: 10.1016/J.MSEC.2015.09.065.

[99] Bender, S., Goellner, J., Heyn, A., Boese, E. (n.d.). Corrosion and corrosion testing of magnesium alloys, Doi: 10.1002/maco.200704091.

[100] Zhao, M.-C., Schmutz, P., Brunner, S., Liu, M., Song, G., Atrens, A. (2009). An exploratory study of the corrosion of Mg alloys during interrupted salt spray testing, *Corros. Sci.*, 51(6), pp. 1277–92, Doi: 10.1016/J.CORSCI.2009.03.014.

[101] Arrabal, R., Pardo, A., Merino, M.C., Mohedano, M., Casajús, P., Paucar, K., Garcés, G. (2012). Effect of Nd on the corrosion behaviour of AM50 and AZ91D magnesium alloys in 3.5 wt.% NaCl solution, *Corros. Sci.*, 55, pp. 301–12, Doi: 10.1016/J.CORSCI.2011.10.033.

[102] Arrabal, R., Matykina, E., Pardo, A., Merino, M.C., Paucar, K., Mohedano, M., Casajús, P. (2012). Corrosion behaviour of AZ91D and AM50 magnesium alloys with Nd and Gd additions in humid environments, *Corros. Sci.*, 55, pp. 351–62, Doi: 10.1016/J.CORSCI.2011.10.038.

[103] Wang, Y., Liao, Z., Song, C., Zhang, H. (2013). Influence of Nd on microstructure and bio-corrosion resistance of Mg-Zn-Mn-Ca alloy, *Rare Met. Mater. Eng.*, 42 (4), pp. 661–6, Doi: 10.1016/S1875-5372(13)60052-1.

[104] Mert, F., Blawert, C., Kainer, K.U., Hort, N. (2012). Influence of cerium additions on the corrosion behaviour of high pressure die cast AM50 alloy, *Corros. Sci.*, 65, pp. 145–51, Doi: 10.1016/J.CORSCI.2012.08.011.

[105] Pekguleryuz, M.O., Kaya, A.A. (2003). Creep resistant magnesium alloys for powertrain applications, *Adv. Eng. Mater.*, 5(12), pp. 866–78, Doi: 10.1002/adem.200300403.

[106] Luo, A.A. (2004). Recent magnesium alloy development for elevated temperature applications, *Int. Mater. Rev.*, 49(1), pp. 13–30, Doi: 10.1179/095066004225010497.

[107] Zhang, J., Liu, S., Leng, Z., Zhang, M., Meng, J., Wu, R. (2011). Microstructures and mechanical properties of heat-resistant HPDC Mg–4Al-based alloys containing cheap misch metal, *Mater. Sci. Eng. A*, 528(6), pp. 2670–7, Doi: 10.1016/J.MSEA.2010.12.031.

[108] Anyanwu, I.A., Gokan, Y., Suzuki, A., Kamado, S., Kojima, Y., Takeda, S., Ishida, T. (2004). Effect of substituting cerium-rich mischmetal with lanthanum on high temperature properties of die-cast Mg–Zn–Al–Ca–RE alloys, *Mater. Sci. Eng. A*, 380(1–2), pp. 93–9, Doi: 10.1016/J.MSEA.2004.03.039.

[109] Zhang, J., Zhang, M., Meng, J., Wu, R., Tang, D. (2010). Microstructures and mechanical properties of heat-resistant high-pressure die-cast Mg–4Al–XLa–0.3Mn (x = 1, 2, 4, 6) alloys, *Mater. Sci. Eng. A*, 527(10–11), pp. 2527–37, Doi: 10.1016/J.MSEA.2009.12.048.

[110] Guangyin, Y., Manping, L., Wenjiang, D., Inoue, A. (2003). Mechanical properties and microstructure of Mg-Al-Zn-Si-base alloy, *Mater. Trans.*, 44(4), pp. 458–62, Doi: 10.2320/matertrans.44.458.

[111] Druschitz, A.P., Showalter, E.R., McNeill, J.B., White, D.L. (2002). Evaluation of structural and high-temperature magnesium alloys. SAE Technical Papers, SAE International.

[112] Zhu, S., Easton, M.A., Abbott, T.B., Nie, J.-F., Dargusch, M.S., Hort, N., Gibson, M.A. (2015). Evaluation of magnesium die-casting alloys for elevated temperature applications: Microstructure, tensile properties, and creep resistance, *Metall. Mater. Trans. A*, 46(8), pp. 3543–54, Doi: 10.1007/s11661-015-2946-9.

[113] Mayer, H., Papakyriacou, M., Zettl, B., Vacic, S. (2005). Endurance limit and threshold stress intensity of die cast magnesium and aluminium alloys at elevated temperatures, *Int. J. Fatigue*, 27(9), pp. 1076–88, Doi: 10.1016/J.IJFATIGUE.2005.02.002.

[114] Berkmortel, J.J., Hu, H., Kearns, J.E., Allison, J.E. (2000). Die castability assessment of magnesium alloys for high temperature applications: Part 1 of 2, *SAE Trans.*, 109, pp. 574–81, Doi: 10.2307/44643878.

[115] Minárik, P., Král, R., Janeček, M. (2013). Effect of ECAP processing on corrosion resistance of AE21 and AE42 magnesium alloys, *Appl. Surf. Sci.*, 281, pp. 44–8, Doi: 10.1016/J.APSUSC.2012.12.096.

[116] Minárik, P., Jablonská, E., Král, R., Lipov, J., Ruml, T., Blawert, C., Hadzima, B., Chmelík, F. (2017). Effect of equal channel angular pressing on in vitro degradation of LAE442 magnesium alloy, *Mater. Sci. Eng. C*, 73, pp. 736–42, Doi: 10.1016/J.MSEC.2016.12.120.

[117] Minárik, P., Král, R., Janeček, M., Chmelík, F., Hadzima, B. (2015). Evolution of corrosion resistance in the LAE442 magnesium alloy processed by ECAP, 128, Doi: 10.12693/APhysPolA.128.772.

[118] Leeflang, M.A., Dzwonczyk, J.S., Zhou, J., Duszczyk, J. (2011). Long-term biodegradation and associated hydrogen evolution of duplex-structured Mg–Li–Al–(RE) alloys and their mechanical properties, *Mater. Sci. Eng. B*, 176(20), pp. 1741–5, Doi: 10.1016/J.MSEB.2011.08.002.

[119] Di Mario, C., Griffiths, H., Goktekin, O., Peeters, N., Verbist, J., Bosiers, M., Deloose, K., Heublein, B., Rohde, R., Kasese, V., Ilsley, C., Erbel, R. (2004). Drug-eluting bioabsorbable magnesium stent, *J. Interv. Cardiol.*, 17(6), pp. 391–5, Doi: 10.1111/j.1540-8183.2004.04081.x.

[120] Witte, F., Fischer, J., Nellesen, J., Crostack, H.A., Kaese, V., Pisch, A., Beckmann, F., Windhagen, H. (2006). In vitro and in vivo corrosion

measurements of magnesium alloys, *Biomaterials*, 27(7), pp. 1013–8, Doi: 10.1016/ j.biomaterials.2005.07.037.

[121] Nordlien, J.H., Nisancioglu, K., Ono, S., Masuko, N. (1997). Morphology and structure of water-formed oxides on ternary MgAl alloys, *J. Electrochem. Soc.*, 144(2), p. 461, Doi: 10.1149/1.1837432.

[122] Mueller, W.-D., Fernández Lorenzo de Mele, M., Nascimento, M.L., Zeddies, M. (2009). Degradation of magnesium and its alloys: Dependence on the composition of the synthetic biological media, *J. Biomed. Mater. Res. Part A*, 90A(2), pp. 487– 95, Doi: 10.1002/jbm.a.32106.

[123] Witte, F., Fischer, J., Nellesen, J., Vogt, C., Vogt, J., Donath, T., Beckmann, F. (2010). In vivo corrosion and corrosion protection of magnesium alloy LAE442, *Acta Biomater.*, 6(5), pp. 1792–9, Doi: 10.1016/J.ACTBIO.2009.10.012.

[124] Wolters, L., Besdo, S., Angrisani, N., Wriggers, P., Hering, B., Seitz, J.-M., Reifenrath, J. (2015). Degradation behaviour of LAE442-based plate–screw-systems in an in vitro bone model, *Mater. Sci. Eng. C*, 49, pp. 305–15, Doi: 10.1016/J.MSEC.2015.01.019.

[125] Ullmann, B., Reifenrath, J., Seitz, J.-M., Bormann, D., Meyer-Lindenberg, A. (2013). Influence of the grain size on the *in vivo* degradation behaviour of the magnesium alloy LAE442, *Proc. Inst. Mech. Eng. Part H J. Eng. Med.*, 227(3), pp. 317–26, Doi: 10.1177/0954411912471495.

[126] Angrisani, N., Reifenrath, J., Zimmermann, F., Eifler, R., Meyer-Lindenberg, A., Vano-Herrera, K., Vogt, C. (2016). Biocompatibility and degradation of LAE442-based magnesium alloys after implantation of up to 3.5 years in a rabbit model, *Acta Biomater.*, 44, pp. 355–65, Doi: 10.1016/J.ACTBIO. 2016.08.002.

[127] Krause, A., von der Höh, N., Bormann, D., Krause, C., Bach, F.-W., Windhagen, H., Meyer-Lindenberg, A. (2010). Degradation behaviour and mechanical properties of magnesium implants in rabbit tibiae, *J. Mater. Sci.*, 45(3), pp. 624–32, Doi: 10.1007/s10853-009-3936-3.

[128] Rössig, C., Angrisani, N., Helmecke, P., Besdo, S., Seitz, J.-M., Welke, B., Fedchenko, N., Kock, H., Reifenrath, J. (2015). In vivo evaluation of a magnesium-based degradable intramedullary nailing system in a sheep model, *Acta Biomater.*, 25, pp. 369–83, Doi: 10.1016/j.actbio.2015.07.025.

[129] Krämer, M., Schilling, M., Eifler, R., Hering, B., Reifenrath, J., Besdo, S., Windhagen, H., Willbold, E., Weizbauer, A. (2016). Corrosion behavior, biocompatibility and biomechanical stability of a prototype magnesium-based biodegradable intramedullary nailing system, *Mater. Sci. Eng. C*, 59, pp. 129–35, Doi: 10.1016/J.MSEC.2015.10.006.

[130] Bracht, K., Angrisani, N., Seitz, J.-M., Eifler, R., Weizbauer, A., Reifenrath, J. (2015). The influence of storage and heat treatment on a magnesium-based implant material: An in vitro and in vivo study, *Biomed. Eng. Online*, 14(1), p. 92, Doi: 10.1186/s12938-015-0091-8.

[131] Khabale, D., Wani, M.F. (2017). Tribological characterization of AZ91 and AE42 magnesium alloys in fretting contact, *J. Tribol.*, 140(1), p. 011604, Doi: 10.1115/ 1.4036922.

[132] Nascimento, M.L., Fleck, C., Müller, W.D., Löhe, D. (2006). Electrochemical characterisation of magnesium and wrought magnesium alloys, *Int. J. Mater. Res.*, 97(11), pp. 1586–93, Doi: 10.3139/146.101425.

[133] Rosalbino, F., Angelini, E., De Negri, S., Saccone, A., Delfino, S. (2006). Electrochemical behaviour assessment of novel Mg-rich Mg–Al–RE alloys (RE = Ce, Er), *Intermetallics*, 14(12), pp. 1487–92, Doi: 10.1016/J.INTERMET.2006.01.056.

[134] Liu, N., Wang, J., Wu, Y., Wang, L. (2008). Electrochemical corrosion behavior of cast Mg–Al–RE–Mn alloys in NaCl solution, *J. Mater. Sci.*, 43(8), pp. 2550–4, Doi: 10.1007/s10853-008-2447-y.

[135] Zhou, W.R., Zheng, Y.F., Leeflang, M.A., Zhou, J. (2013). Mechanical property, biocorrosion and in vitro biocompatibility evaluations of Mg–Li–(Al)–(RE) alloys for future cardiovascular stent application, *Acta Biomater.*, 9(10), pp. 8488–98, Doi: 10.1016/J.ACTBIO.2013.01.032.

[136] Hampp, C., Angrisani, N., Reifenrath, J., Bormann, D., Seitz, J.-M., Meyer-Lindenberg, A. (2013). Evaluation of the biocompatibility of two magnesium alloys as degradable implant materials in comparison to titanium as non-resorbable material in the rabbit, *Mater. Sci. Eng. C*, 33(1), pp. 317–26, Doi: 10.1016/J.MSEC.2012.08.046.

[137] Reifenrath, J., Krause, A., Bormann, D., von Rechenberg, B., Windhagen, H., Meyer-Lindenberg, A. (2010). Profound differences in the in-vivo-degradation and biocompatibility of two very similar rare-earth containing Mg-alloys in a rabbit model. Massive Unterschiede im in-vivo-Degradationsverhalten und in der Biokompatibilität zweier sehr ähnlicher Seltene-Er, *Materwiss. Werksttech.*, 41 (12), pp. 1054–61, Doi: 10.1002/mawe.201000709.

[138] Ullmann, B., Reifenrath, J., Dziuba, D., Seitz, J.-M., Bormann, D., Meyer-Lindenberg, A. (2011). In vivo degradation behavior of the magnesium alloy LANd442 in rabbit tibiae, *Materials* (Basel)., 4(12), pp. 2197–218, Doi: 10.3390/ma4122197.

[139] Brandt, E.G., Hellgren, M., Brinck, T., Bergman, T., Edholm, O. (2009). Molecular dynamics study of zinc binding to cysteines in a peptide mimic of the alcohol dehydrogenase structural zinc site, *Phys. Chem. Chem. Phys.*, 11(6), pp. 975–83, Doi: 10.1039/b815482a.

[140] Prasad, A.S. (2008). Zinc in human health: Effect of zinc on immune cells, *Mol. Med.*, 14(5–6), pp. 353–7, Doi: 10.2119/2008-00033.Prasad.

[141] Avedesian, M.M., Baker, H., ASM International. Handbook Committee. (1999). *Magnesium and magnesium alloys*, Cleveland, OH, ASM International.

[142] Zhang, L.-N., Hou, Z.-T., Ye, X., Xu, Z.-B., Bai, X.-L., Shang, P. (2013). The effect of selected alloying element additions on properties of Mg-based alloy as bioimplants: A literature review, *Front. Mater. Sci.*, 7(3), pp. 227–36, Doi: 10.1007/s11706-013-0210-z.

[143] Hanawalt, J.D., Nelson, C.E., Peloubet, J.A. (1942). Corrosion studies of magnesium and its alloys, *Trans AIME*, 47, pp. 273–99.

[144] Shaw, B.A. (2003). Corrosion resistance of magnesium alloys. In *ASM handbook*, Vol. 13A, Pennsylvania State University, pp. 692–6.

[145] Zhang, S., Zhang, X., Zhao, C., Li, J., Song, Y., Xie, C., Tao, H., Zhang, Y., He, Y., Jiang, Y., Bian, Y. (2010). Research on an Mg-Zn alloy as a degradable biomaterial, *Acta Biomater.*, 6(2), pp. 626–40, Doi: 10.1016/J.ACTBIO.2009.06.028.

[146] Persaud-Sharma, D., McGoron, A. (2012). Biodegradable magnesium alloys: A review of material development and applications, *J. Biomim. Biomater. Tissue Eng.*, 12(2011), pp. 25–39, Doi: 10.4028/www.scientific.net/JBBTE.12.25.

[147] Koh, J.Y., Choi, D.W. (1994). Zinc toxicity on cultured cortical neurons: Involvement of N-methyl-D-aspartate receptors, *Neuroscience*, 60(4), pp. 1049–57.

[148] Borovanský, J., Riley, P.A. (1989). Cytotoxicity of zinc in vitro, *Chem. Biol. Interact.*, 69(2–3), pp. 279–91, Doi: 10.1016/0009-2797(89)90085-9.

[149] Bothwell, D.N., Mair, E.A., Cable, B.B. (2003). Chronic ingestion of a zinc-based penny, *Pediatrics*, 111(3), pp. 689–91.

[150] Ku, C.-H., Pioletti, D.P., Browne, M., Gregson, P.J. (2002). Effect of different Ti–6Al–4V surface treatments on osteoblasts behaviour, *Biomaterials*, 23(6), pp. 1447–54, Doi: 10.1016/S0142-9612(01)00266-6.

[151] Nakamura, Y., Tsumura, Y., Tonogai, Y., Shibata, T., Ito, Y. (1997). Differences in behavior among the chlorides of seven rare earth elements administered intravenously to rats, *Fundam. Appl. Toxicol.*, 37(2), pp. 106–16, Doi: 10.1006/FAAT.1997.2322.

[152] Peng, Q., Li, X., Ma, N., Liu, R., Zhang, H. (2012). Effects of backward extrusion on mechanical and degradation properties of Mg–Zn biomaterial, *J. Mech. Behav. Biomed. Mater.*, 10, pp. 128–37, Doi: 10.1016/J.JMBBM.2012.02.024.

[153] Cai, S., Lei, T., Li, N., Feng, F. (2012). Effects of Zn on microstructure, mechanical properties and corrosion behavior of Mg–Zn alloys, *Mater. Sci. Eng. C*, 32(8), pp. 2570–7, Doi: 10.1016/J.MSEC.2012.07.042.

[154] Cheng, M., Chen, J., Yan, H., Su, B., Yu, Z., Xia, W., Gong, X. (2017). Effects of minor Sr addition on microstructure, mechanical and bio-corrosion properties of the Mg-5Zn based alloy system, *J. Alloys Compd.*, 691, pp. 95–102, Doi: 10.1016/J.JALLCOM.2016.08.164.

[155] Li, H., Peng, Q., Li, X., Li, K., Han, Z., Fang, D. (2014). Microstructures, mechanical and cytocompatibility of degradable Mg–Zn based orthopedic biomaterials, *Mater. Des.*, 58, pp. 43–51, Doi: 10.1016/J.MATDES.2014.01.031.

[156] Suganthi, R.V., Elayaraja, K., Joshy, M.I.A., Chandra, V.S., Girija, E.K., Kalkura, S.N. (2011). Fibrous growth of strontium substituted hydroxyapatite and its drug release, *Mater. Sci. Eng. C*, 31(3), pp. 593–9, Doi: 10.1016/J.MSEC.2010.11.025.

[157] Zhang, W., Shen, Y., Pan, H., Lin, K., Liu, X., Darvell, B.W., Lu, W.W., Chang, J., Deng, L., Wang, D., Huang, W. (2011). Effects of strontium in modified biomaterials, *Acta Biomater.*, 7(2), pp. 800–8, Doi: 10.1016/J.ACTBIO.2010.08.031.

[158] Guan, R., Cipriano, A.F., Zhao, Z., Lock, J., Tie, D., Zhao, T., Cui, T., Liu, H. (2013). Development and evaluation of a magnesium–zinc–strontium alloy for biomedical applications – Alloy processing, microstructure, mechanical properties, and biodegradation, *Mater. Sci. Eng. C*, 33(7), pp. 3661–9, Doi: 10.1016/J.MSEC.2013.04.054.

[159] Zhang, B.P., Wang, Y., Geng, L. (2011). Research on Mg-Zn-Ca alloy as degradable biomaterial. In *Biomaterials – Physics and chemistry*, Ed. R. Pignatello, London, UK, InTech, pp. 183–204.

[160] Liu, X., Shan, D., Song, Y., Han, E. (2010). Effects of heat treatment on corrosion behaviors of Mg-3Zn magnesium alloy, *Trans. Nonferrous Met. Soc. China*, 20(7), pp. 1345–50, Doi: 10.1016/S1003-6326(09)60302-2.

[161] Zhang, S., Li, J., Song, Y., Zhao, C., Zhang, X., Xie, C., Zhang, Y., Tao, H., He, Y., Jiang, Y., Bian, Y. (2009). In vitro degradation, hemolysis and MC3T3-E1 cell adhesion of biodegradable Mg–Zn alloy, *Mater. Sci. Eng. C*, 29(6), pp. 1907–12, Doi: 10.1016/J.MSEC.2009.03.001.

[162] Cui, T., Guan, R., Ma, X., Qin, H., Song, F. (2018). Studies on biocompatibility of Mg-4.0Zn-1.5Sr alloy with coated of the laser surface processing combining alkaline treatment, *IOP Conf. Ser. Mater. Sci. Eng.*, 301(1), p. 012076, Doi: 10.1088/1757-899X/301/1/012076.

[163] United States Pharmacopeial Convention, and United States Pharmacopoeial Convention (1980). *The United States pharmacopeia: The national formulary*, Rockville, MD, United States Pharmacopeial Convention.

[164] Fazel Anvari-Yazdi, A., Tahermanesh, K., Hadavi, S.M.M., Talaei-Khozani, T., Razmkhah, M., Abed, S.M., Mohtasebi, M.S. (2016). Cytotoxicity assessment of

adipose-derived mesenchymal stem cells on synthesized biodegradable Mg-Zn-Ca alloys, *Mater. Sci. Eng. C*, 69, pp. 584–97, Doi: 10.1016/J.MSEC.2016.07.016.

[165] Yan, Y., Kang, Y., Li, D., Yu, K., Xiao, T., Deng, Y., Dai, H., Dai, Y., Xiong, H., Fang, H. (2017). Improvement of the mechanical properties and corrosion resistance of biodegradable β-Ca3(PO4)2/Mg-Zn composites prepared by powder metallurgy: The adding β-Ca3(PO4)2, hot extrusion and aging treatment, *Mater. Sci. Eng. C*, 74, pp. 582–96, Doi: 10.1016/J.MSEC.2016.12.132.

[166] Seyedraoufi, Z.S., Mirdamadi, S. (2015). In vitro biodegradability and biocompatibility of porous Mg-Zn scaffolds coated with nano hydroxyapatite via pulse electrodeposition, *Trans. Nonferrous Met. Soc. China*, 25(12), pp. 4018–27, Doi: 10.1016/S1003-6326(15)64051-1.

[167] Li, J., Song, Y., Zhang, S., Zhao, C., Zhang, F., Zhang, X., Cao, L., Fan, Q., Tang, T. (2010). In vitro responses of human bone marrow stromal cells to a fluoridated hydroxyapatite coated biodegradable Mg–Zn alloy, *Biomaterials*, 31 (22), pp. 5782–8, Doi: 10.1016/J.BIOMATERIALS.2010.04.023.

[168] Li, Z., Chen, M., Li, W., Zheng, H., You, C., Liu, D., Jin, F. (2017). The synergistic effect of trace Sr and Zr on the microstructure and properties of a biodegradable Mg-Zn-Zr-Sr alloy, *J. Alloys Compd.*, 702, pp. 290–302, Doi: 10.1016/J.JALLCOM.2017.01.178.

[169] Nguyen, T.Y., Cipriano, A.F., Guan, R.-G., Zhao, Z.-Y., Liu, H. (2015). *In vitro* interactions of blood, platelet, and fibroblast with biodegradable magnesium-zinc-strontium alloys, *J. Biomed. Mater. Res. Part A*, 103(9), pp. 2974–86, Doi: 10.1002/jbm.a.35429.

[170] Cipriano, A.F., Sallee, A., Tayoba, M., Cortez Alcaraz, M.C., Lin, A., Guan, R.-G., Zhao, Z.-Y., Liu, H. (2017). Cytocompatibility and early inflammatory response of human endothelial cells in direct culture with Mg-Zn-Sr alloys, *Acta Biomater.*, 48, pp. 499–520, Doi: 10.1016/J.ACTBIO.2016.10.020.

[171] He, Y., Tao, H., Zhang, Y., Jiang, Y., Zhang, S., Zhao, C., Li, J., Zhang, B., Song, Y., Zhang, X. (2009). Biocompatibility of bio-Mg-Zn alloy within bone with heart, liver, kidney and spleen, *Sci. Bull.*, 54(3), pp. 484–91, Doi: 10.1007/s11434-009-0080-z.

[172] Wang, Z., Yan, J., Li, J., Zheng, Q., Wang, Z., Zhang, X., Zhang, S. (2012). Effects of biodegradable Mg–6Zn alloy extracts on apoptosis of intestinal epithelial cells, *Mater. Sci. Eng. B*, 177(4), pp. 388–93, Doi: 10.1016/J.MSEB.2012.01.002.

[173] StJohn, D.H., Qian, M., Easton, M.A., Cao, P., Hildebrand, Z. (2005). Grain refinement of magnesium alloys, *Metall. Mater. Trans. A*, 36(7), pp. 1669–79, Doi: 10.1007/s11661-005-0030-6.

[174] Lee, D.B.N., Roberts, M., Bluchel, C.G., Odell, R.A. (2010). Zirconium: Biomedical and nephrological applications, *ASAIO J.*, 56(6), pp. 550–6, Doi: 10.1097/MAT.0b013e3181e73f20.

[175] Saldana, L., Mendezvilas, A., Jiang, L., Multigner, M., Gonzalezcarrasco, J., Perezprado, M., Gonzalezmartin, M., Munuera, L., Vilaboa, N. (2007). In vitro biocompatibility of an ultrafine grained zirconium, *Biomaterials*, 28(30), pp. 4343–54, Doi: 10.1016/j.biomaterials.2007.06.015.

[176] Emsley, J. (2001). *Nature's building blocks: An A-Z guide to the elements*, Oxford, UK, Oxford University Press.

[177] Jamesh, M.I., Wu, G., Zhao, Y., McKenzie, D.R., Bilek, M.M.M., Chu, P.K. (2014). Effects of zirconium and oxygen plasma ion implantation on the corrosion behavior of ZK60 Mg alloy in simulated body fluids, *Corros. Sci.*, 82, pp. 7–26, Doi: 10.1016/J.CORSCI.2013.11.044.

[178] Zheng, Y.F., Gu, X.N., Witte, F. (2014). Biodegradable metals, *Mater. Sci. Eng. R Reports*, 77, pp. 1–34, Doi: 10.1016/j.mser.2014.01.001.

[179] Volkova, E.F. (2006). Effect of deformation and heat treatment on the structure and properties of magnesium alloys of the Mg-Zn-Zr system, *Met. Sci. Heat Treat.*, 48(11–12), pp. 508–12, Doi: 10.1007/s11041-006-0127-6.

[180] Nair, K.S., Mittal, M.C., Sikand, R., Gupta, A.K., Gupta, A.K. (2008). Effect of extrusion parameters on microstructure and mechanical properties of ZK30 Mg alloy, *Mater. Sci. Technol.*, 24(4), pp. 399–405, Doi: 10.1179/174328408X276224.

[181] Chen, H., Zang, Q., Yu, H., Zhang, J., Jin, Y. (2015). Effect of intermediate annealing on the microstructure and mechanical property of ZK60 magnesium alloy produced by twin roll casting and hot rolling, *Mater. Charact.*, 106, pp. 437–41, Doi: 10.1016/J.MATCHAR.2015.06.015.

[182] Choi, H.Y., Kim, W.J. (2015). Effect of thermal treatment on the bio-corrosion and mechanical properties of ultrafine-grained ZK60 magnesium alloy, *J. Mech. Behav. Biomed. Mater.*, 51, pp. 291–301, Doi: 10.1016/j.jmbbm.2015.07.019.

[183] Yu, H., Hongge, Y., Jihua, C., Bin, S., Yi, Z., Yanjin, S., Zhaojie, M. (2014). Effects of minor Gd addition on microstructures and mechanical properties of the high strain-rate rolled Mg–Zn–Zr alloys, *J. Alloys Compd.*, 586, pp. 757–65, Doi: 10.1016/J.JALLCOM.2013.10.005.

[184] Yin, D.L., Cui, H.L., Qiao, J., Zhang, J.F. (2015). Enhancement of mechanical properties in a Mg–Zn–Zr alloy by equal channel angular pressing at warm temperature, *Mater. Res. Innov.*, 19(sup9), pp. s9-28–s9-31, Doi: 10.1179/1432891715Z.0000000001912.

[185] Mostaed, E., Fabrizi, A., Dellasega, D., Bonollo, F., Vedani, M. (2015). Grain size and texture dependence on mechanical properties, asymmetric behavior and low temperature superplasticity of ZK60 Mg alloy, *Mater. Charact.*, 107, pp. 70–8, Doi: 10.1016/J.MATCHAR.2015.06.009.

[186] Yuan, Y., Ma, A., Jiang, J., Gou, X., Song, D., Yang, D., Jian, W. (2016). High mechanical properties of rolled ZK60 Mg alloy through pre-equal channel angular pressing, *Mechanics*, 22(4), pp. 256–9, Doi: 10.5755/j01.mech.22.4.16161.

[187] Orlov, D., Raab, G., Lamark, T.T., Popov, M., Estrin, Y. (2011). Improvement of mechanical properties of magnesium alloy ZK60 by integrated extrusion and equal channel angular pressing, *Acta Mater.*, 59(1), pp. 375–85, Doi: 10.1016/J.ACTAMAT.2010.09.043.

[188] Ying, T., Huang, J., Zheng, M., Wu, K. (2012). Influence of secondary extrusion on microstructures and mechanical properties of ZK60 Mg alloy processed by extrusion and ECAP, *Trans. Nonferrous Met. Soc. China*, 22(8), pp. 1896–901, Doi: 10.1016/S1003-6326(11)61404-0.

[189] Tao, J., Cheng, Y., Huang, S., Peng, F., Yang, W., Lu, M., Zhang, Z., Jin, X. (2012). Microstructural evolution and mechanical properties of ZK60 magnesium alloy prepared by multi-axial forging during partial remelting, *Trans. Nonferrous Met. Soc. China*, 22, pp. s428–34, Doi: 10.1016/S1003-6326(12)61742-7.

[190] Lin, J., Wang, Q., Peng, L., Roven, H.J. (2009). Microstructure and high tensile ductility of ZK60 magnesium alloy processed by cyclic extrusion and compression, *J. Alloys Compd.*, 476(1–2), pp. 441–5, Doi: 10.1016/J.JALLCOM.2008.09.031.

[191] He, Y., Pan, Q., Qin, Y., Liu, X., Li, W. (2010). Microstructure and mechanical properties of ultrafine grain ZK60 alloy processed by equal channel angular pressing, *J. Mater. Sci.*, 45(6), pp. 1655–62, Doi: 10.1007/s10853-009-4143-y.

[192] Wu, Y.-Z., Yan, H.-G., Chen, J.-H., Du, Y.-G., Zhu, S.-Q., Su, B. (2012). Microstructure and mechanical properties of ZK21 magnesium alloy fabricated by multiple forging at different strain rates, *Mater. Sci. Eng. A*, 556, pp. 164–9, Doi: 10.1016/J.MSEA.2012.06.074.

[193] He, Y., Pan, Q., Qin, Y., Liu, X., Li, W., Chiu, Y., Chen, J.J.J. (2010). Microstructure and mechanical properties of ZK60 alloy processed by two-step equal channel

angular pressing, *J. Alloys Compd.*, 492(1–2), pp. 605–10, Doi: 10.1016/J. JALLCOM.2009.11.192.

[194] Chen, B., Zhang, J. (2015). Microstructure and mechanical properties of ZK60-Er magnesium alloys, *Mater. Sci. Eng. A*, 633, pp. 154–60, Doi: 10.1016/J. MSEA.2015.03.009.

[195] Wang, C., Zhang, Y., Li, D., Mei, H., Zhang, W., Liu, J. (2013). Microstructure evolution and mechanical properties of ZK60 magnesium alloy produced by SSTT and RAP route in semi-solid state, *Trans. Nonferrous Met. Soc. China*, 23 (12), pp. 3621–8, Doi: 10.1016/S1003-6326(13)62909-X.

[196] Jin, W., Fan, J., Zhang, H., Liu, Y., Dong, H., Xu, B. (2015). Microstructure, mechanical properties and static recrystallization behavior of the rolled ZK60 magnesium alloy sheets processed by electropulsing treatment, *J. Alloys Compd.*, 646, pp. 1–9, Doi: 10.1016/J.JALLCOM.2015.04.196.

[197] Markushev, M.V., Nugmanov, D.R., Sitdikov, O., Vinogradov, A. (2018). Structure, texture and strength of Mg-5.8Zn-0.65Zr alloy after hot-to-warm multi-step isothermal forging and isothermal rolling to large strains, *Mater. Sci. Eng. A*, 709, pp. 330–8, Doi: 10.1016/J.MSEA.2017.10.008.

[198] Yuan, Y., Ma, A., Gou, X., Jiang, J., Lu, F., Song, D., Zhu, Y. (2015). Superior mechanical properties of ZK60 mg alloy processed by equal channel angular pressing and rolling, *Mater. Sci. Eng. A*, 630, pp. 45–50, Doi: 10.1016/J. MSEA.2015.02.004.

[199] Kim, W.J., Moon, I.K., Han, S.H. (2012). Ultrafine-grained Mg–Zn–Zr alloy with high strength and high-strain-rate superplasticity, *Mater. Sci. Eng. A*, 538, pp. 374–85, Doi: 10.1016/J.MSEA.2012.01.063.

[200] Chen, X., Le, Q., Wang, X., Liao, Q., Chu, C. (2017). Variable-frequency ultrasonic treatment on microstructure and mechanical properties of ZK60 alloy during large diameter semi-continuous casting, *Metals* (Basel)., 7(5), p. 173, Doi: 10.3390/met7050173.

[201] Dumitru, F.-D., Higuera-Cobos, O.F., Cabrera, J.M. (2014). ZK60 alloy processed by ECAP: Microstructural, physical and mechanical characterization, *Mater. Sci. Eng. A*, 594, pp. 32–9, Doi: 10.1016/J.MSEA.2013.11.050.

[202] Galiyev, A., Kaibyshev, R., Gottstein, G. (2001). Correlation of plastic deformation and dynamic recrystallization in magnesium alloy ZK60, *Acta Mater.*, 49(7), pp. 1199–207, Doi: 10.1016/S1359-6454(01)00020-9.

[203] Yuan, Y., Ma, A., Gou, X., Jiang, J., Arhin, G., Song, D., Liu, H. (2016). Effect of heat treatment and deformation temperature on the mechanical properties of ECAP processed ZK60 magnesium alloy, *Mater. Sci. Eng. A*, 677, pp. 125–32, Doi: 10.1016/J.MSEA.2016.09.037.

[204] Goodman, S.B., Davidson, J.A., Fornasier, V.L., Mishra, A.K. (1993). Histological response to cylinders of a low modulus titanium alloy (Ti-13Nb-13Zr) and a wear resistant zirconium alloy (Zr-2.5Nb) implanted in the rabbit tibia, *J. Appl. Biomater.*, 4(4), pp. 331–9, Doi: 10.1002/jab.770040407.

[205] Song, G., StJohn, D. (2002). The effect of zirconium grain refinement on the corrosion behaviour of magnesium-rare earth alloy MEZ, *J. Light Met.*, 2(1), pp. 1–16, Doi: 10.1016/S1471-5317(02)00008-1.

[206] Hong, D., Saha, P., Chou, D.-T., Lee, B., Collins, B.E., Tan, Z., Dong, Z., Kumta, P.N. (2013). In vitro degradation and cytotoxicity response of Mg–4% Zn–0.5% Zr (ZK40) alloy as a potential biodegradable material, *Acta Biomater.*, 9 (10), pp. 8534–47, Doi: 10.1016/J.ACTBIO.2013.07.001.

[207] Zhang, S., Bi, Y., Li, J., Wang, Z., Yan, J., Song, J., Sheng, H., Guo, H., Li, Y. (2017). Biodegradation behavior of magnesium and ZK60 alloy in artificial urine

and rat models, *Bioact. Mater.*, 2(2), pp. 53–62, Doi: 10.1016/J.BIOACTMAT. 2017.03.004.

[208] Gao, J., Wu, S., Qiao, L., Wang, Y. (2008). Corrosion behavior of Mg and Mg-Zn alloys in simulated body fluid, *Trans. Nonferrous Met. Soc. China*, 18(3), pp. 588–92, Doi: 10.1016/S1003-6326(08)60102-8.

[209] Huan, Z.G., Leeflang, M.A., Zhou, J., Fratila-Apachitei, L.E., Duszczyk, J. (2010). In vitro degradation behavior and cytocompatibility of Mg-Zn-Zr alloys, *J. Mater. Sci. Mater. Med.*, 21(9), pp. 2623–35, Doi: 10.1007/s10856-010-4111-8.

[210] Li, T., Zhang, H., He, Y., Wen, N., Wang, X. (2015). Microstructure, mechanical properties and in vitro degradation behavior of a novel biodegradable Mg–1.5Zn–0.6Zr–0.2Sc alloy, *J. Mater. Sci. Technol.*, 31(7), pp. 744–50, Doi: 10.1016/J. JMST.2015.02.001.

[211] MSIT®, M.S.I.T. (n.d.). *Mg-Zn-Zr (Magnesium – Zinc – Zirconium). Light metal systems. Part 4*, Berlin/Heidelberg, Springer-Verlag, pp. 1–6.

[212] Mostaed, E., Hashempour, M., Fabrizi, A., Dellasega, D., Bestetti, M., Bonollo, F., Vedani, M. (2014). Microstructure, texture evolution, mechanical properties and corrosion behavior of ECAP processed ZK60 magnesium alloy for biodegradable applications, *J. Mech. Behav. Biomed. Mater.*, 37, pp. 307–22, Doi: 10.1016/J.JMBBM.2014.05.024.

[213] Schroeder, H.A., Balassa, J.J. (1966). Abnormal trace metals in man: Zirconium, *J. Chronic Dis.*, 19(5), pp. 573–86, Doi: 10.1016/0021-9681(66)90095-6.

[214] Zhang, Q., Lin, X., Qi, Z., Tan, L., Yang, K., Hu, Z., Wang, Y. (2013). Magnesium alloy for repair of lateral tibial plateau defect in minipig model, *J. Mater. Sci. Technol.*, 29(6), pp. 539–44, Doi: 10.1016/J.JMST.2013.03.003.

[215] Van der Stok, J., Van Lieshout, E.M.M., El-Massoudi, Y., Van Kralingen, G.H., Patka, P. (2011). Bone substitutes in the Netherlands – A systematic literature review, *Acta Biomater.*, 7(2), pp. 739–50, Doi: 10.1016/J.ACTBIO.2010.07.035.

[216] Peters, C.L., Hines, J.L., Bachus, K.N., Craig, M.A., Bloebaum, R.D. (2006). Biological effects of calcium sulfate as a bone graft substitute in ovine metaphyseal defects, *J. Biomed. Mater. Res. Part A*, 76A(3), pp. 456–62, Doi: 10.1002/jbm. a.30569.

[217] Lin, X., Tan, L., Wang, Q., Zhang, G., Zhang, B., Yang, K. (2013). In vivo degradation and tissue compatibility of ZK60 magnesium alloy with micro-arc oxidation coating in a transcortical model, *Mater. Sci. Eng. C*, 33(7), pp. 3881–8, Doi: 10.1016/J.MSEC.2013.05.023.

[218] Pan, Y.K., Chen, C.Z., Wang, D.G., Zhao, T.G. (2014). Improvement of corrosion and biological properties of microarc oxidized coatings on Mg–Zn–Zr alloy by optimizing negative power density parameters, *Colloids Surf. B Biointerfaces*, 113, pp. 421–8, Doi: 10.1016/J.COLSURFB.2013.09.044.

[219] Pan, Y.K., Chen, C.Z., Wang, D.G., Zhao, T.G. (2013). Effects of phosphates on microstructure and bioactivity of micro-arc oxidized calcium phosphate coatings on Mg–Zn–Zr magnesium alloy, *Colloids Surf. B Biointerfaces*, 109, pp. 1–9, Doi: 10.1016/J.COLSURFB.2013.03.026.

[220] Jin, W., Hao, Q., Peng, X., Chu, P.K. (2016). Enhanced corrosion resistance and biocompatibilty of PMMA-coated ZK60 magnesium alloy, *Mater. Lett.*, 173, pp. 178–81, Doi: 10.1016/J.MATLET.2016.03.071.

[221] Ye, X., Chen, M., Yang, M., Wei, J., Liu, D. (2010). In vitro corrosion resistance and cytocompatibility of nano-hydroxyapatite reinforced Mg–Zn–Zr composites, *J. Mater. Sci. Mater. Med.*, 21(4), pp. 1321–8, Doi: 10.1007/s10856-009-3954-3.

[222] Zheng, H.R., Li, Z., You, C., Liu, D.B., Chen, M.F. (2017). Effects of MgO modified β-TCP nanoparticles on the microstructure and properties of β-TCP/Mg-Zn-Zr

composites, *Bioact. Mater.*, 2(1), pp. 1–9, Doi: 10.1016/J.BIOACTMAT. 2016.12.004.

[223] Yamasaki, Y., Yoshida, Y., Okazaki, M., Shimazu, A., Kubo, T., Akagawa, Y., Uchida, T. (2003). Action of FGMgCO3Ap-collagen composite in promoting bone formation, *Biomaterials*, 24(27), pp. 4913–20.

[224] Zreiqat, H., Howlett, C.R., Zannettino, A., Evans, P., Schulze-Tanzil, G., Knabe, C., Shakibaei, M. (2002). Mechanisms of magnesium-stimulated adhesion of osteoblastic cells to commonly used orthopaedic implants, *J. Biomed. Mater. Res.*, 62(2), pp. 175–84, Doi: 10.1002/jbm.10270.

[225] Jin, W., Wang, G., Lin, Z., Feng, H., Li, W., Peng, X., Qasim, A.M., Chu, P.K. (2017). Corrosion resistance and cytocompatibility of tantalum-surface-functionalized biomedical ZK60 Mg alloy, *Corros. Sci.*, 114, pp. 45–56, Doi: 10.1016/J.CORSCI.2016.10.021.

[226] Ilich, J.Z., Kerstetter, J.E. (n.d.). Nutrition in bone health revisited: A story beyond calcium, *J. Am. Coll. Nutr.*, 19(6), pp. 715–37.

[227] Serre, C.M., Papillard, M., Chavassieux, P., Voegel, J.C., Boivin, G. (1998). Influence of magnesium substitution on a collagen-apatite biomaterial on the production of a calcifying matrix by human osteoblasts, *J. Biomed. Mater. Res.*, 42(4), pp. 626–33.

[228] Zhang, B., Wang, Y., Geng, L., Lu, C. (2012). Effects of calcium on texture and mechanical properties of hot-extruded Mg–Zn–Ca alloys, *Mater. Sci. Eng. A*, 539, pp. 56–60, Doi: 10.1016/J.MSEA.2012.01.030.

[229] Bettles, C.J., Gibson, M.A., Venkatesan, K. (2004). Enhanced age-hardening behaviour in Mg–4 wt.% Zn micro-alloyed with Ca, *Scr. Mater.*, 51(3), pp. 193–7, Doi: 10.1016/J.SCRIPTAMAT.2004.04.020.

[230] Geng, L., Zhang, B.P., Li, A.B., Dong, C.C. (2009). Microstructure and mechanical properties of Mg–4.0Zn–0.5Ca alloy, *Mater. Lett.*, 63(5), pp. 557–9, Doi: 10.1016/J.MATLET.2008.11.044.

[231] Yang, L., Zhang, E. (2009). Biocorrosion behavior of magnesium alloy in different simulated fluids for biomedical application, *Mater. Sci. Eng. C*, 29(5), pp. 1691–6, Doi: 10.1016/J.MSEC.2009.01.014.

[232] Cha, P.-R., Han, H.-S., Yang, G.-F., Kim, Y.-C., Hong, K.-H., Lee, S.-C., Jung, J.-Y., Ahn, J.-P., Kim, Y.-Y., Cho, S.-Y., Byun, J.Y., Lee, K.-S., Yang, S.-J., Seok, H.-K. (2013). Biodegradability engineering of biodegradable Mg alloys: Tailoring the electrochemical properties and microstructure of constituent phases, *Sci. Rep.*, 3(1), p. 2367, Doi: 10.1038/srep02367.

[233] Li, Z., Gu, X., Lou, S., Zheng, Y. (2008). The development of binary Mg–Ca alloys for use as biodegradable materials within bone, *Biomaterials*, 29(10), pp. 1329–44, Doi: 10.1016/J.BIOMATERIALS.2007.12.021.

[234] Zander, D., Zumdick, N.A. (2015). Influence of Ca and Zn on the microstructure and corrosion of biodegradable Mg–Ca–Zn alloys, *Corros. Sci.*, 93, pp. 222–33, Doi: 10.1016/J.CORSCI.2015.01.027.

[235] Bakhsheshi-Rad, H.R., Idris, M.H., Abdul-Kadir, M.R., Ourdjini, A., Medraj, M., Daroonparvar, M., Hamzah, E. (2014). Mechanical and bio-corrosion properties of quaternary Mg–Ca–Mn–Zn alloys compared with binary Mg–Ca alloys, *Mater. Des.*, 53, pp. 283–92, Doi: 10.1016/J. MATDES.2013.06.055.

[236] Du, Y.Z., Qiao, X.G., Zheng, M.Y., Wang, D.B., Wu, K., Golovin, I.S. (2016). Effect of microalloying with Ca on the microstructure and mechanical properties of Mg-6 mass%Zn alloys, *Mater. Des.*, 98, pp. 285–93, Doi: 10.1016/J. MATDES.2016.03.025.

[237] Du, Y., Zheng, M., Qiao, X., Wang, D., Peng, W., Wu, K., Jiang, B. (2016). Improving microstructure and mechanical properties in Mg–6 mass% Zn alloys by combined addition of Ca and Ce, *Mater. Sci. Eng. A*, 656, pp. 67–74, Doi: 10.1016/J.MSEA.2016.01.034.

[238] Oh-ishi, K., Mendis, C.L., Homma, T., Kamado, S., Ohkubo, T., Hono, K. (2009). Bimodally grained microstructure development during hot extrusion of Mg–2.4 Zn–0.1 Ag–0.1 Ca–0.16 Zr (at.%) alloys, *Acta Mater.*, 57(18), pp. 5593–604, Doi: 10.1016/J.ACTAMAT.2009.07.057.

[239] Mendis, C.L., Oh-ishi, K., Kawamura, Y., Honma, T., Kamado, S., Hono, K. (2009). Precipitation-hardenable Mg–2.4Zn–0.1Ag–0.1Ca–0.16Zr (at.%) wrought magnesium alloy, *Acta Mater.*, 57(3), pp. 749–60, Doi: 10.1016/J. ACTAMAT.2008.10.033.

[240] Tong, L.B., Zheng, M.Y., Hu, X.S., Wu, K., Xu, S.W., Kamado, S., Kojima, Y. (2010). Influence of ECAP routes on microstructure and mechanical properties of Mg–Zn–Ca alloy, *Mater. Sci. Eng. A*, 527(16–17), pp. 4250–6, Doi: 10.1016/J. MSEA.2010.03.062.

[241] Tong, L.B., Zheng, M.Y., Chang, H., Hu, X.S., Wu, K., Xu, S.W., Kamado, S., Kojima, Y. (2009). Microstructure and mechanical properties of Mg–Zn–Ca alloy processed by equal channel angular pressing, *Mater. Sci. Eng. A*, 523(1–2), pp. 289–94, Doi: 10.1016/J.MSEA.2009.06.021.

[242] Li, C., Sun, H., Li, X., Zhang, J., Fang, W., Tan, Z. (2015). Microstructure, texture and mechanical properties of Mg-3.0Zn-0.2Ca alloys fabricated by extrusion at various temperatures, *J. Alloys Compd.*, 652, pp. 122–31, Doi: 10.1016/J. JALLCOM.2015.08.215.

[243] Tong, L.B., Zheng, M.Y., Xu, S.W., Hu, X.S., Wu, K., Kamado, S., Wang, G.J., Lv, X.Y. (2010). Room-temperature compressive deformation behavior of Mg–Zn–Ca alloy processed by equal channel angular pressing, *Mater. Sci. Eng. A*, 528 (2), pp. 672–9, Doi: 10.1016/J.MSEA.2010.09.041.

[244] Dobroň, P., Drozdenko, D., Olejňák, J., Hegedüs, M., Horváth, K., Veselý, J., Bohlen, J., Letzig, D. (2018). Compressive yield stress improvement using thermo-mechanical treatment of extruded Mg-Zn-Ca alloy, *Mater. Sci. Eng. A*, 730, pp. 401–9, Doi: 10.1016/J.MSEA.2018.06.026.

[245] Jiang, M.G., Xu, C., Nakata, T., Yan, H., Chen, R.S., Kamado, S. (2016). Development of dilute Mg–Zn–Ca–Mn alloy with high performance via extrusion, *J. Alloys Compd.*, 668, pp. 13–21, Doi: 10.1016/J.JALLCOM.2016.01.195.

[246] Tong, L.B., Zhang, J.B., Zhang, Q.X., Jiang, Z.H., Xu, C., Kamado, S., Zhang, D. P., Meng, J., Cheng, L.R., Zhang, H.J. (2016). Effect of warm rolling on the microstructure, texture and mechanical properties of extruded Mg–Zn–Ca–Ce/La alloy, *Mater. Charact.*, 115, pp. 1–7, Doi: 10.1016/J.MATCHAR.2016.03.012.

[247] Homma, T., Mendis, C.L., Hono, K., Kamado, S. (2010). Effect of Zr addition on the mechanical properties of as-extruded Mg–Zn–Ca–Zr alloys, *Mater. Sci. Eng. A*, 527(9), pp. 2356–62, Doi: 10.1016/J.MSEA.2009.12.024.

[248] Luo, J., Yan, H., Zheng, N., Chen, R.-S. (2016). Effects of zinc and calcium concentration on the microstructure and mechanical properties of hot-rolled Mg–Zn–Ca sheets, *Acta Metall. Sin. (English Lett.)*, 29(2), pp. 205–16, Doi: 10.1007/ s40195-016-0378-1.

[249] Wang, G., Huang, G., Chen, X., Deng, Q., Tang, A., Jiang, B., Pan, F. (2017). Effects of Zn addition on the mechanical properties and texture of extruded Mg-Zn-Ca-Ce magnesium alloy sheets, *Mater. Sci. Eng. A*, 705, pp. 46–54, Doi: 10.1016/J.MSEA.2017.08.036.

[250] Jiang, M.G., Xu, C., Nakata, T., Yan, H., Chen, R.S., Kamado, S. (2016). High-speed extrusion of dilute Mg-Zn-Ca-Mn alloys and its effect on microstructure,

texture and mechanical properties, *Mater. Sci. Eng. A*, 678, pp. 329–38, Doi: 10.1016/J.MSEA.2016.10.007.

[251] Du, Y., Zheng, M., Jiang, B. (2018). Comparison of microstructure and mechanical properties of Mg-Zn microalloyed with Ca or Ce, *Vacuum*, 151, pp. 221–5, Doi: 10.1016/J.VACUUM.2018.02.029.

[252] Zeng, Z.R., Bian, M.Z., Xu, S.W., Davies, C.H.J., Birbilis, N., Nie, J.F. (2016). Effects of dilute additions of Zn and Ca on ductility of magnesium alloy sheet, *Mater. Sci. Eng. A*, 674, pp. 459–71, Doi: 10.1016/J.MSEA.2016.07.049.

[253] Du, Y.Z., Qiao, X.G., Zheng, M.Y., Wu, K., Xu, S.W. (2015). The microstructure, texture and mechanical properties of extruded Mg–5.3Zn–0.2Ca–0.5Ce (wt%) alloy, *Mater. Sci. Eng. A*, 620, pp. 164–71, Doi: 10.1016/J.MSEA.2014.10.028.

[254] Ding, H., Shi, X., Wang, Y., Cheng, G., Kamado, S. (2015). Texture weakening and ductility variation of Mg–2Zn alloy with CA or RE addition, *Mater. Sci. Eng. A*, 645, pp. 196–204, Doi: 10.1016/J.MSEA.2015.08.025.

[255] Massalski, T.B., Okamoto, H. (Hiroaki)., ASM International. (1990). *Binary alloy phase diagrams*, Cleveland, OH, ASM International.

[256] Zhang, B., Hou, Y., Wang, X., Wang, Y., Geng, L. (2011). Mechanical properties, degradation performance and cytotoxicity of Mg–Zn–Ca biomedical alloys with different compositions, *Mater. Sci. Eng. C*, 31(8), pp. 1667–73, Doi: 10.1016/J.MSEC.2011.07.015.

[257] Bakhsheshi-Rad, H.R., Abdul-Kadir, M.R., Idris, M.H., Farahany, S. (2012). Relationship between the corrosion behavior and the thermal characteristics and microstructure of Mg–0.5Ca–xZn alloys, *Corros. Sci.*, 64, pp. 184–97, Doi: 10.1016/J.CORSCI.2012.07.015.

[258] Hänzi, A.C., Sologubenko, A.S., Gunde, P., Schinhammer, M., Uggowitzer, P.J. (2012). Design considerations for achieving simultaneously high-strength and highly ductile magnesium alloys, *Philos. Mag. Lett.*, 92(9), pp. 417–27, Doi: 10.1080/09500839.2012.657701.

[259] Wilson, D.V., Chapman, J.A. (1963). Effects of preferred orientation on the grain size dependence of yield strength in metals, *Philos. Mag.*, 8(93), pp. 1543–51, Doi: 10.1080/14786436308207317.

[260] Barnett, M.R., Keshavarz, Z., Beer, A.G., Atwell, D. (2004). Influence of grain size on the compressive deformation of wrought Mg–3Al–1Zn, *Acta Mater.*, 52 (17), pp. 5093–103, Doi: 10.1016/J.ACTAMAT.2004.07.015.

[261] Aung, N.N., Zhou, W. (2010). Effect of grain size and twins on corrosion behaviour of AZ31B magnesium alloy, *Corros. Sci.*, 52(2), pp. 589–94, Doi: 10.1016/J.CORSCI.2009.10.018.

[262] Zhang, C.Z., Zhu, S.J., Wang, L.G., Guo, R.M., Yue, G.C., Guan, S.K. (2016). Microstructures and degradation mechanism in simulated body fluid of biomedical Mg–Zn–Ca alloy processed by high pressure torsion, *Mater. Des.*, 96, pp. 54–62, Doi: 10.1016/J.MATDES.2016.01.072.

[263] Gao, J.H., Guan, S.K., Ren, Z.W., Sun, Y.F., Zhu, S.J., Wang, B. (2011). Homogeneous corrosion of high pressure torsion treated Mg–Zn–Ca alloy in simulated body fluid, *Mater. Lett.*, 65(4), pp. 691–3, Doi: 10.1016/J.MATLET.2010.11.015.

[264] Renkema, K.Y., Alexander, R.T., Bindels, R.J., Hoenderop, J.G. (2008). Calcium and phosphate homeostasis: Concerted interplay of new regulators, *Ann. Med.*, 40 (2), pp. 82–91, Doi: 10.1080/07853890701689645.

[265] Cipriano, A.F., Sallee, A., Guan, R.-G., Zhao, Z.-Y., Tayoba, M., Sanchez, J., Liu, H. (2015). Investigation of magnesium–zinc–calcium alloys and bone marrow derived mesenchymal stem cell response in direct culture, *Acta Biomater.*, 12, pp. 298–321, Doi: 10.1016/J.ACTBIO.2014.10.018.

[266] Zhao, Y.F., Si, J.J., Song, J.G., Hui, X.D. (2014). High strength Mg–Zn–Ca alloys prepared by atomization and hot pressing process, *Mater. Lett.*, 118, pp. 55–8, Doi: 10.1016/J.MATLET.2013.12.053.

[267] Gu, X., Zheng, Y., Zhong, S., Xi, T., Wang, J., Wang, W. (2010). Corrosion of, and cellular responses to Mg–Zn–Ca bulk metallic glasses, *Biomaterials*, 31(6), pp. 1093–103, Doi: 10.1016/J.BIOMATERIALS.2009.11.015.

[268] Matias, T.B., Roche, V., Nogueira, R.P., Asato, G.H., Kiminami, C.S., Bolfarini, C., Botta, W.J., Jorge, A.M. (2016). Mg-Zn-Ca amorphous alloys for application as temporary implant: Effect of Zn content on the mechanical and corrosion properties, *Mater. Des.*, 110, pp. 188–95, Doi: 10.1016/J.MATDES.2016.07.148.

[269] Inoue, A. (2000). Stabilization of metallic supercooled liquid and bulk amorphous alloys, *Acta Mater.*, 48(1), pp. 279–306, Doi: 10.1016/S1359-6454(99)00300-6.

[270] Ma, E., Xu, J. (2009). The glass window of opportunities, *Nat. Mater.*, 8(11), pp. 855–7, Doi: 10.1038/nmat2550.

[271] Hui, X., Dong, W., Chen, G.L., Yao, K.F. (2007). Formation, microstructure and properties of long-period order structure reinforced Mg-based bulk metallic glass composites, *Acta Mater.*, 55(3), pp. 907–20, Doi: 10.1016/J.ACTAMAT.2006.09.012.

[272] Li, Q.-F., Weng, H.-R., Suo, Z.-Y., Ren, Y.-L., Yuan, X.-G., Qiu, K.-Q. (2008). Microstructure and mechanical properties of bulk Mg–Zn–Ca amorphous alloys and amorphous matrix composites, *Mater. Sci. Eng. A*, 487(1–2), pp. 301–8, Doi: 10.1016/J.MSEA.2007.10.027.

[273] Wang, J., Huang, S., Wei, Y., Guo, S., Fusheng, P. (2013). Enhanced mechanical properties and corrosion resistance of a Mg–Zn–Ca bulk metallic glass composite by Fe particle addition, *Mater. Lett.*, 91, pp. 311–4, Doi: 10.1016/J.MATLET.2012.09.098.

[274] Shanthi, M., Gupta, M., Jarfors, A.E.W., Tan, M.J. (2011). Synthesis, characterization and mechanical properties of nano alumina particulate reinforced magnesium based bulk metallic glass composites, *Mater. Sci. Eng. A*, 528(18), pp. 6045–50, Doi: 10.1016/J.MSEA.2011.03.103.

[275] Gebert, A., Wolff, U., John, A., Eckert, J., Schultz, L. (2001). Stability of the bulk glass-forming Mg65Y10Cu25 alloy in aqueous electrolytes, *Mater. Sci. Eng. A*, 299(1–2), pp. 125–35, Doi: 10.1016/S0921-5093(00)01401-5.

[276] Wang, W.H., Dong, C., Shek, C.H. (2004). Bulk metallic glasses, *Mater. Sci. Eng. R Reports*, 44(2–3), pp. 45–89, Doi: 10.1016/J.MSER.2004.03.001.

[277] Nowosielski, R., Cesarz-Andraczke, K. (2018). Impact of Zn and Ca on dissolution rate, mechanical properties and GFA of resorbable Mg–Zn–Ca metallic glasses, *Arch. Civ. Mech. Eng.*, 18(1), pp. 1–11, Doi: 10.1016/J.ACME.2017.05.009.

[278] Wong, P.-C., Tsai, P.-H., Li, T.-H., Cheng, C.-K., Jang, J.S.C., Huang, J.C. (2017). Degradation behavior and mechanical strength of Mg-Zn-Ca bulk metallic glass composites with Ti particles as biodegradable materials, *J. Alloys Compd.*, 699, pp. 914–20, Doi: 10.1016/J.JALLCOM.2017.01.010.

[279] Allodi, I., Udina, E., Navarro, X. (2012). Specificity of peripheral nerve regeneration: Interactions at the axon level, *Prog. Neurobiol.*, 98(1), pp. 16–37, Doi: 10.1016/J.PNEUROBIO.2012.05.005.

[280] Sutherland, D.J., Pujic, Z., Goodhill, G.J. (2014). Calcium signaling in axon guidance, *Trends Neurosci.*, 37(8), pp. 424–32, Doi: 10.1016/J.TINS.2014.05.008.

[281] Gower-Winter, S.D., Levenson, C.W. (2012). Zinc in the central nervous system: From molecules to behavior, *BioFactors*, 38(3), pp. 186–93, Doi: 10.1002/biof.1012.

[282] Zhang, X.F., Coughlan, A., O'Shea, H., Towler, M.R., Kehoe, S., Boyd, D. (2012). Experimental composite guidance conduits for peripheral nerve repair: An evaluation of ion release, *Mater. Sci. Eng. C*, 32(6), pp. 1654–63, Doi: 10.1016/J.MSEC.2012.04.058.

[283] Li, H., He, W., Pang, S., Liaw, P.K., Zhang, T. (2016). In vitro responses of bone-forming MC3T3-E1 pre-osteoblasts to biodegradable Mg-based bulk metallic glasses, *Mater. Sci. Eng. C*, 68, pp. 632–41, Doi: 10.1016/J.MSEC.2016.06.022.

[284] Buha, J. (2008). Reduced temperature (22–100°C) ageing of an Mg–Zn alloy, *Mater. Sci. Eng. A*, 492(1–2), pp. 11–19, Doi: 10.1016/j.msea.2008.02.038.

[285] Zhang, Z., Couture, A., Luo, A. (1998). An investigation of the properties of Mg-Zn-Al alloys, *Scr. Mater.*, 39(1), pp. 45–53, Doi: 10.1016/S1359-6462(98)00122-5.

[286] Moreno, I.P., Nandy, T.K., Jones, J.W., Allison, J.E., Pollock, T.M. (2003). Microstructural stability and creep of rare-earth containing magnesium alloys, *Scr. Mater.*, 48(8), pp. 1029–34, Doi: 10.1016/S1359-6462(02)00595-X.

[287] Boehlert, C.J., Knittel, K. (2006). The microstructure, tensile properties, and creep behavior of Mg-Zn alloys containing 0–4.4 wt.% Zn, *Mater. Sci. Eng. A*, 417, pp. 315–21, Doi: 10.1016/j.msea.2005.11.006.

[288] Zhang, D., Shi, G., Dai, Q., Yuan, W., Duan, H. (2008). Microstructures and mechanical properties of high strength Mg-Zn-Mn alloy, *Trans. Nonferrous Met. Soc. China*, 18, pp. s59–63, Doi: 10.1016/S1003-6326(10)60175-6.

[289] Yuan, J., Zhang, K., Li, T., Li, X., Li, Y., Ma, M., Luo, P., Luo, G., Hao, Y. (2012). Anisotropy of thermal conductivity and mechanical properties in Mg–5Zn–1Mn alloy, *Mater. Des.*, 40, pp. 257–61, Doi: 10.1016/J.MATDES.2012.03.046.

[290] Khan, S.A., Miyashita, Y., Mutoh, Y., Sajuri, Z. Bin. (2006). Influence of Mn content on mechanical properties and fatigue behavior of extruded Mg alloys, *Mater. Sci. Eng. A*, 420(1–2), pp. 315–21, Doi: 10.1016/J.MSEA.2006.01.091.

[291] Cho, D.H., Lee, B.W., Park, J.Y., Cho, K.M., Park, I.M. (2017). Effect of Mn addition on corrosion properties of biodegradable Mg-4Zn-0.5Ca-xMn alloys, *J. Alloys Compd.*, 695, pp. 1166–74, Doi: 10.1016/J.JALLCOM.2016.10.244.

[292] Cho, D.H., Nam, J.H., Lee, B.W., Cho, K.M., Park, I.M. (2016). Effect of Mn addition on grain refinement of biodegradable Mg4Zn0.5Ca alloy, *J. Alloys Compd.*, 676, pp. 461–8, Doi: 10.1016/J.JALLCOM.2016.03.182.

[293] Xu, L., Yu, G., Zhang, E., Pan, F., Yang, K. (2007). In vivo corrosion behavior of Mg-Mn-Zn alloy for bone implant application, *J. Biomed. Mater. Res. Part A*, 83A(3), pp. 703–11, Doi: 10.1002/jbm.a.31273.

[294] Helsen, J.A., Breme, H.J. (1998). *Metals as biomaterials*, Danvers, MA, John Wiley & Sons.

[295] Zhang, D., Qi, F., Lan, W., Shi, G., Zhao, X. (2011). Effects of Ce addition on microstructure and mechanical properties of Mg-6Zn-1Mn alloy, *Trans. Nonferrous Met. Soc. China*, 21(4), pp. 703–10, Doi: 10.1016/S1003-6326(11)60769-3.

[296] Nie, J.F. (2003). Effects of precipitate shape and orientation on dispersion strengthening in magnesium alloys, *Scr. Mater.*, 48(8), pp. 1009–15, Doi: 10.1016/S1359-6462(02)00497-9.

[297] Hu, G., Zhang, D., Tang, T., Shen, X., Jiang, L., Xu, J., Pan, F. (2015). Effects of Nd addition on microstructure and mechanical properties of Mg–6Zn–1Mn–4Sn alloy, *Mater. Sci. Eng. A*, 634, pp. 5–13, Doi: 10.1016/J.MSEA.2015.03.040.

[298] Yang, J., Peng, J., Li, M., Nyberg, E.A., Pan, F.-S. (2017). Effects of Ca addition on the mechanical properties and corrosion behavior of ZM21 wrought alloys, *Acta Metall. Sin. (English Lett.)*, 30(1), pp. 53–65, Doi: 10.1007/s40195-016-0492-0.

[299] Park, S.H., Jung, J.-G., Yoon, J., You, B.S. (2015). Influence of Sn addition on the microstructure and mechanical properties of extruded Mg–8Al–2Zn alloy, *Mater. Sci. Eng. A*, 626, pp. 128–35, Doi: 10.1016/J.MSEA.2014.12.039.

[300] Jiang, Z., Jiang, B., Yang, H., Yang, Q., Dai, J., Pan, F. (2015). Influence of the Al2Ca phase on microstructure and mechanical properties of Mg–Al–Ca alloys, *J. Alloys Compd.*, 647, pp. 357–63, Doi: 10.1016/J.JALLCOM.2015.06.060.

[301] Pan, F., Mao, J., Zhang, G., Tang, A., She, J. (2016). Development of high-strength, low-cost wrought Mg–2.0 mass% Zn alloy with high Mn content, *Prog. Nat. Sci. Mater. Int.*, 26(6), pp. 630–5, Doi: 10.1016/J.PNSC.2016.11.016.

[302] Qi, F., Zhang, D., Zhang, X., Pan, F. (2014). Effect of Y addition on microstructure and mechanical properties of Mg–Zn–Mn alloy, *Trans. Nonferrous Met. Soc. China*, 24(5), pp. 1352–64, Doi: 10.1016/S1003-6326(14)63199-X.

[303] Zhang, D., Shi, G., Zhao, X., Qi, F. (2011). Microstructure evolution and mechanical properties of Mg-x%Zn-1%Mn (x=4, 5, 6, 7, 8, 9) wrought magnesium alloys, *Trans. Nonferrous Met. Soc. China*, 21(1), pp. 15–25, Doi: 10.1016/S1003-6326(11)60672-9.

[304] Yin, D., Zhang, E., Zeng, S. (2008). Effect of Zn on mechanical property and corrosion property of extruded Mg-Zn-Mn alloy, *Trans. Nonferrous Met. Soc. China*, 18(4), pp. 763–8, Doi: 10.1016/S1003-6326(08)60131-4.

[305] Rosalbino, F., De Negri, S., Scavino, G., Saccone, A. (2013). Microstructure and *in vitro* degradation performance of Mg-Zn-Mn alloys for biomedical application, *J. Biomed. Mater. Res. Part A*, 101A(3), pp. 704–11, Doi: 10.1002/jbm.a.34368.

[306] Zhang, Y., Li, J., Li, J. (2018). Effects of microstructure transformation on mechanical properties, corrosion behaviors of Mg-Zn-Mn-Ca alloys in simulated body fluid, *J. Mech. Behav. Biomed. Mater.*, 80, pp. 246–57, Doi: 10.1016/J.JMBBM.2018.01.028.

[307] Zeng, R.-C., Sun, L., Zheng, Y.-F., Cui, H.-Z., Han, E.-H. (2014). Corrosion and characterisation of dual phase Mg–Li–Ca alloy in Hank's solution: The influence of microstructural features, *Corros. Sci.*, 79, pp. 69–82, Doi: 10.1016/J.CORSCI.2013.10.028.

[308] Pu, Z., Song, G.-L., Yang, S., Outeiro, J.C., Dillon, O.W., Puleo, D.A., Jawahir, I.S. (2012). Grain refined and basal textured surface produced by burnishing for improved corrosion performance of AZ31B Mg alloy, *Corros. Sci.*, 57, pp. 192–201, Doi: 10.1016/J.CORSCI.2011.12.018.

[309] Zhang, Y., Li, J., Lai, H., Xu, Y. (2018). Effect of homogenization on microstructure characteristics, corrosion and biocompatibility of Mg-Zn-Mn-xCa alloys, *Materials* (Basel)., 11(2), p. 227, Doi: 10.3390/ma11020227.

[310] Zhang, E., Yin, D., Xu, L., Yang, L., Yang, K. (2009). Microstructure, mechanical and corrosion properties and biocompatibility of Mg–Zn–Mn alloys for biomedical application, *Mater. Sci. Eng. C*, 29(3), pp. 987–93, Doi: 10.1016/J.MSEC.2008.08.024.

[311] Lee, J.Y., Kim, D.H., Lim, H.K., Kim, D.H. (2005). Effects of Zn/Y ratio on microstructure and mechanical properties of Mg-Zn-Y alloys, *Mater. Lett.*, 59(29–30), pp. 3801–5, Doi: 10.1016/J.MATLET.2005.06.052.

[312] Luo, Z., Zhang, S., Tang, Y., Zhao, D. (1993). Quasicrystals in as-cast Mg-Zn-RE alloys, *Scr. Metall. Mater.*, 28(12), pp. 1513–8, Doi: 10.1016/0956-716X(93)90584-F.

[313] Xu, D.K., Liu, L., Xu, Y.B., Han, E.H. (2006). The influence of element Y on the mechanical properties of the as-extruded Mg–Zn–Y–Zr alloys, *J. Alloys Compd.*, 426(1–2), pp. 155–61, Doi: 10.1016/J.JALLCOM.2006.02.035.

[314] Tsai, A.-P., Murakami, Y., Niikura, A. (2000). The Zn-Mg-Y phase diagram involving quasicrystals, *Philos. Mag. A*, 80(5), pp. 1043–54, Doi: 10.1080/01418610008212098.

[315] Singh, A., Tsai, A.P., Nakamura, M., Watanabe, M., Kato, A. (2003). Nanoprecipitates of icosahedral phase in quasicrystal-strengthened Mg-Zn-Y alloys, *Philos. Mag. Lett.*, 83(9), pp. 543–51, Doi: 10.1080/09500830310001597027.

[316] Gröbner, J., Kozlov, A., Fang, X.Y., Geng, J., Nie, J.F., Schmid-Fetzer, R. (2012). Phase equilibria and transformations in ternary Mg-rich Mg–Y–Zn alloys, *Acta Mater.*, 60(17), pp. 5948–62, Doi: 10.1016/J.ACTAMAT.2012.05.035.

[317] Yang, K., Zhang, J., Zong, X., Wang, W., Xu, C., Cheng, W., Nie, K. (2016). Effect of microalloying with boron on the microstructure and mechanical properties of Mg–Zn–Y–Mn alloy, *Mater. Sci. Eng. A*, 669, pp. 340–3, Doi: 10.1016/J. MSEA.2016.05.095.

[318] Wang, J., Zhang, J., Zong, X., Xu, C., You, Z., Nie, K. (2015). Effects of Ca on the formation of LPSO phase and mechanical properties of Mg-Zn-Y-Mn alloy, *Mater. Sci. Eng. A*, 648, pp. 37–40, Doi: 10.1016/J.MSEA.2015.09.046.

[319] Xu, D.K., Tang, W.N., Liu, L., Xu, Y.B., Han, E.H. (2007). Effect of Y concentration on the microstructure and mechanical properties of as-cast Mg–Zn–Y–Zr alloys, *J. Alloys Compd.*, 432(1–2), pp. 129–34, Doi: 10.1016/J. JALLCOM.2006.05.123.

[320] Wang, J., Gao, S., Song, P., Huang, X., Shi, Z., Pan, F. (2011). Effects of phase composition on the mechanical properties and damping capacities of as-extruded Mg–Zn–Y–Zr alloys, *J. Alloys Compd.*, 509(34), pp. 8567–72, Doi: 10.1016/J. JALLCOM.2011.06.017.

[321] Wang, J., Song, P., Gao, S., Huang, X., Shi, Z., Pan, F. (2011). Effects of Zn on the microstructure, mechanical properties, and damping capacity of Mg–Zn–Y–Zr alloys, *Mater. Sci. Eng. A*, 528(18), pp. 5914–20, Doi: 10.1016/J.MSEA. 2011.04.002.

[322] Wang, J., Song, P., Gao, S., Wei, Y., Pan, F. (2012). Influence of Y on the phase composition and mechanical properties of as-extruded Mg–Zn–Y–Zr magnesium alloys, *J. Mater. Sci.*, 47(4), pp. 2005–10, Doi: 10.1007/s10853-011-5998-2.

[323] Gao, X., Nie, J.F. (2008). Enhanced precipitation-hardening in Mg–Gd alloys containing Ag and Zn, *Scr. Mater.*, 58(8), pp. 619–22, Doi: 10.1016/J.SCRIPTAMAT. 2007.11.022.

[324] Xu, C., Zheng, M.Y., Wu, K., Wang, E.D., Fan, G.H., Xu, S.W., Kamado, S., Liu, X.D., Wang, G.J., Lv, X.Y. (2013). Influence of rolling temperature on the microstructure and mechanical properties of Mg–Gd–Y–Zn–Zr alloy sheets, *Mater. Sci. Eng. A*, 559, pp. 615–22, Doi: 10.1016/J.MSEA.2012.08.151.

[325] Zhu, Y.M., Weyland, M., Morton, A.J., Oh-ishi, K., Hono, K., Nie, J.F. (2009). The building block of long-period structures in Mg–RE–Zn alloys, *Scr. Mater.*, 60 (11), pp. 980–3, Doi: 10.1016/J.SCRIPTAMAT.2009.02.029.

[326] Yamasaki, M., Matsushita, M., Hagihara, K., Izuno, H., Abe, E., Kawamura, Y. (2014). Highly ordered 10H-type long-period stacking order phase in a Mg–Zn–Y ternary alloy, *Scr. Mater.*, 78–79, pp. 13–16, Doi: 10.1016/J.SCRIPTAMAT. 2014.01.013.

[327] Abe, E., Ono, A., Itoi, T., Yamasaki, M., Kawamura, Y. (2011). Polytypes of long-period stacking structures synchronized with chemical order in a dilute Mg–Zn–Y alloy, *Philos. Mag. Lett.*, 91(10), pp. 690–6, Doi: 10.1080/09500839.2011.609149.

[328] Zhu, Y.M., Morton, A.J., Nie, J.F. (2010). The 18R and 14H long-period stacking ordered structures in Mg–Y–Zn alloys, *Acta Mater.*, 58(8), pp. 2936–47, Doi: 10.1016/J.ACTAMAT.2010.01.022.

[329] Yin, D.D., Wang, Q.D., Gao, Y., Chen, C.J., Zheng, J. (2011). Effects of heat treatments on microstructure and mechanical properties of Mg–11Y–5Gd–2Zn–0.5Zr (wt.%) alloy, *J. Alloys Compd.*, 509(5), pp. 1696–704, Doi: 10.1016/J. JALLCOM.2010.09.194.

[330] Itoi, T., Inazawa, T., Yamasaki, M., Kawamura, Y., Hirohashi, M. (2013). Microstructure and mechanical properties of Mg ZnY alloy sheet prepared by hot-rolling, *Mater. Sci. Eng. A*, 560, pp. 216–23, Doi: 10.1016/J.MSEA.2012.09.059.

[331] Yan, B., Dong, X., Ma, R., Chen, S., Pan, Z., Ling, H. (2014). Effects of heat treatment on microstructure, mechanical properties and damping capacity of Mg–Zn–Y–Zr alloy, *Mater. Sci. Eng. A*, 594, pp. 168–77, Doi: 10.1016/J.MSEA.2013.11.019.

[332] Oñorbe, E., Garcés, G., Pérez, P., Adeva, P. (2012). Effect of the LPSO volume fraction on the microstructure and mechanical properties of Mg–Y2X –Zn X alloys, *J. Mater. Sci.*, 47(2), pp. 1085–93, Doi: 10.1007/s10853-011-5899-4.

[333] Kim, J., Kawamura, Y. (2013). Influence of rare earth elements on microstructure and mechanical properties of Mg97Zn1Y1RE1 alloys, *Mater. Sci. Eng. A*, 573, pp. 62–6, Doi: 10.1016/J.MSEA.2012.12.087.

[334] Xu, Z., Smith, C., Chen, S., Sankar, J. (2011). Development and microstructural characterizations of Mg–Zn–Ca alloys for biomedical applications, *Mater. Sci. Eng. B*, 176(20), pp. 1660–5, Doi: 10.1016/J.MSEB.2011.06.008.

[335] Zhang, H., Feng, J., Zhu, W., Liu, C., Wu, D., Yang, W., Gu, J. (2000). Rare-earth element distribution characteristics of biological chains in rare-earth element-high background regions and their implications, *Biol. Trace Elem. Res.*, 73(1), pp. 19–28, Doi: 10.1385/BTER:73:1:19.

[336] Ding, Y., Wen, C., Hodgson, P., Li, Y. (2014). Effects of alloying elements on the corrosion behavior and biocompatibility of biodegradable magnesium alloys: A review, *J. Mater. Chem. B*, 2(14), pp. 1912–33, Doi: 10.1039/C3TB21746A.

[337] Loos, A., Rohde, R., Haverich, A., Barlach, S. (n.d.). In vitro and in vivo biocompatibility testing of absorbable metal stents, Doi: 10.1002/masy.200750715.

[338] Lee, J.Y., Lim, H.K., Kim, D.H., Kim, W.T., Kim, D.H. (2008). Effect of icosahedral phase particles on the texture evolution in Mg–Zn–Y alloys, *Mater. Sci. Eng. A*, 491(1–2), pp. 349–55, Doi: 10.1016/J.MSEA.2008.02.010.

[339] Singh, A., Watanabe, M., Kato, A., Tsai, A.P. (2005). Twinning and the orientation relationships of icosahedral phase with the magnesium matrix, *Acta Mater.*, 53(17), pp. 4733–42, Doi: 10.1016/J.ACTAMAT.2005.06.026.

[340] Singh, A., Tsai, A.P. (2005). A new orientation relationship OR4 of icosahedral phase with magnesium matrix in Mg–Zn–Y alloys, *Scr. Mater.*, 53(9), pp. 1083–7, Doi: 10.1016/J.SCRIPTAMAT.2005.06.041.

[341] Xu, D.K., Han, E.H. (2012). Effects of icosahedral phase formation on the microstructure and mechanical improvement of Mg alloys: A review, *Prog. Nat. Sci. Mater. Int.*, 22(5), pp. 364–85, Doi: 10.1016/J.PNSC.2012.09.005.

[342] Inoue, A., Kawamura, Y., Matsushita, M., Hayashi, K., Koike, J. (2001). Novel hexagonal structure and ultrahigh strength of magnesium solid solution in the Mg–Zn–Y system, *J. Mater. Res.*, 16(07), pp. 1894–900, Doi: 10.1557/JMR.2001.0260.

[343] Yamasaki, M., Hashimoto, K., Hagihara, K., Kawamura, Y. (2011). Effect of multimodal microstructure evolution on mechanical properties of Mg–Zn–Y extruded alloy, *Acta Mater.*, 59(9), pp. 3646–58, Doi: 10.1016/J.ACTAMAT.2011.02.038.

[344] Shao, X.H., Yang, Z.Q., Ma, X.L. (2010). Strengthening and toughening mechanisms in Mg–Zn–Y alloy with a long period stacking ordered structure, *Acta Mater.*, 58(14), pp. 4760–71, Doi: 10.1016/J.ACTAMAT.2010.05.012.

[345] Cheng, P., Zhao, Y., Lu, R., Hou, H. (2018). Effect of the morphology of long-period stacking ordered phase on mechanical properties and corrosion behavior of cast Mg-Zn-Y-Ti alloy, *J. Alloys Compd.*, 764, pp. 226–38, Doi: 10.1016/J.JALLCOM.2018.06.056.

[346] Feng, H., Liu, H., Cao, H., Yang, Y., Xu, Y., Guan, J. (2015). Effect of precipitates on mechanical and damping properties of Mg–Zn–Y–Nd alloys, *Mater. Sci. Eng. A*, 639, pp. 1–7, Doi: 10.1016/J.MSEA.2015.04.092.

[347] Chen, T.J., Zhang, D.H., Wang, W., Ma, Y., Hao, Y. (2014). Effects of Zn content on microstructures and mechanical properties of Mg–Zn–RE–Sn–Zr–Ca alloys, *Mater. Sci. Eng. A*, 607, pp. 17–27, Doi: 10.1016/J.MSEA.2014.03.111.

[348] Xu, D.K., Tang, W.N., Liu, L., Xu, Y.B., Han, E.H. (2008). Effect of W-phase on the mechanical properties of as-cast Mg–Zn–Y–Zr alloys, *J. Alloys Compd.*, 461 (1–2), pp. 248–52, Doi: 10.1016/J.JALLCOM.2007.07.096.

[349] Mishra, R.K., Gupta, A.K., Rao, P.R., Sachdev, A.K., Kumar, A.M., Luo, A.A. (2008). Influence of cerium on the texture and ductility of magnesium extrusions, *Scr. Mater.*, 59(5), pp. 562–5, Doi: 10.1016/J.SCRIPTAMAT.2008.05.019.

[350] Ball, E.A., Prangnell, P.B. (1994). Tensile-compressive yield asymmetries in high strength wrought magnesium alloys, *Scr. Metall. Mater.*, 31(2), pp. 111–6, Doi: 10.1016/0956-716X(94)90159-7.

[351] Mackenzie, L.W.F., Pekguleryuz, M.O. (2008). The recrystallization and texture of magnesium–zinc–cerium alloys, *Scr. Mater.*, 59(6), pp. 665–8, Doi: 10.1016/J.SCRIPTAMAT.2008.05.021.

[352] Stanford, N., Barnett, M.R. (2008). The origin of "rare earth" texture development in extruded Mg-based alloys and its effect on tensile ductility, *Mater. Sci. Eng. A*, 496(1–2), pp. 399–408, Doi: 10.1016/J.MSEA.2008.05.045.

[353] Wu, T., Jin, L., Wu, W.X., Gao, L., Wang, J., Zhang, Z.Y., Dong, J. (2013). Improved ductility of Mg–Zn–Ce alloy by hot pack-rolling, *Mater. Sci. Eng. A*, 584, pp. 97–102, Doi: 10.1016/J.MSEA.2013.07.011.

[354] Pérez, P., Onofre, E., Cabeza, S., Llorente, I., del Valle, J.A., García-Alonso, M.C., Adeva, P., Escudero, M.L. (2013). Corrosion behaviour of Mg–Zn–Y–Mischmetal alloys in phosphate buffer saline solution, *Corros. Sci.*, 69, pp. 226–35, Doi: 10.1016/J.CORSCI.2012.12.007.

[355] Zhao, X., Shi, L., Xu, J. (2013). Mg–Zn–Y alloys with long-period stacking ordered structure: In vitro assessments of biodegradation behavior, *Mater. Sci. Eng. C*, 33(7), pp. 3627–37, Doi: 10.1016/J.MSEC.2013.04.051.

[356] Li, X., Mao, P., Wang, F., Wang, Z., Liu, Z., Zhou, L. (2018). Effect of heat treatments on mechanical properties and corrosion behavior of MgY3Zn2Al magnesium alloy, Doi: 10.1088/2053-1591/aad9da.

[357] Wang, J., Wang, L., Guan, S., Zhu, S., Ren, C., Hou, S. (2010). Microstructure and corrosion properties of as sub-rapid solidification Mg–Zn–Y–Nd alloy in dynamic simulated body fluid for vascular stent application, *J. Mater. Sci. Mater. Med.*, 21(7), pp. 2001–8, Doi: 10.1007/s10856-010-4063-z.

[358] Zhang, E., He, W., Du, H., Yang, K. (2008). Microstructure, mechanical properties and corrosion properties of Mg–Zn–Y alloys with low Zn content, *Mater. Sci. Eng. A*, 488(1–2), pp. 102–11, Doi: 10.1016/J.MSEA.2007.10.056.

[359] Chiu, C., Lu, C.-T., Chen, S.-H., Ou, K.-L. (2017). Effect of hydroxyapatite on the mechanical properties and corrosion behavior of Mg-Zn-Y alloy, *Materials* (Basel)., 10(8), p. 855, Doi: 10.3390/ma10080855.

[360] Shi, F., Yu, Y., Guo, X., Zhang, Z., Li, Y. (2009). Corrosion behavior of as-cast Mg68Zn28Y4 alloy with I-phase, *Trans. Nonferrous Met. Soc. China*, 19(5), pp. 1093–7, Doi: 10.1016/S1003-6326(08)60412-4.

[361] Wang, S.D., Xu, D.K., Chen, X.B., Han, E.H., Dong, C. (2015). Effect of heat treatment on the corrosion resistance and mechanical properties of an as-forged Mg–Zn–Y–Zr alloy, *Corros. Sci.*, 92, pp. 228–36, Doi: 10.1016/J.CORSCI.2014.12.008.

[362] Wu, Q., Zhu, S., Wang, L., Liu, Q., Yue, G., Wang, J., Guan, S. (2012). The microstructure and properties of cyclic extrusion compression treated Mg–Zn–Y–Nd alloy for vascular stent application, *J. Mech. Behav. Biomed. Mater.*, 8, pp. 1–7, Doi: 10.1016/J.JMBBM.2011.12.011.

[363] Feyerabend, F., Fischer, J., Holtz, J., Willumeit, R., Drücker, H., Vogt, C., Hort, N. (2010). Evaluation of short-term effects of rare earth and other elements used in magnesium alloys on primary cells and cell lines, *Acta Biomater.*, 6(5), pp. 1834–42, Doi: 10.1016/J.ACTBIO.2009.09.024.

[364] Rim, K.T., Koo, K.H., Park, J.S. (2013). Toxicological evaluations of rare earths and their health impacts to workers: A literature review, *Saf. Health Work*, 4(1), pp. 12–26, Doi: 10.5491/SHAW.2013.4.1.12.

[365] Liu, X., Zhen, Z., Liu, J., Xi, T., Zheng, Y., Guan, S., Zheng, Y., Cheng, Y. (2015). Multifunctional MgF2/polydopamine coating on Mg alloy for vascular stent application, *J. Mater. Sci. Technol.*, 31(7), pp. 733–43, Doi: 10.1016/J.JMST.2015.02.002.

[366] Chen, L., Li, J., Chang, J., Jin, S., Wu, D., Yan, H., Wang, X., Guan, S. (2018). Mg-Zn-Y-Nd coated with citric acid and dopamine by layer-by-layer self-assembly to improve surface biocompatibility, *Sci. China Technol. Sci.*, 61(8), pp. 1228–37, Doi: 10.1007/s11431-017-9190-2.

[367] Liu, J., Wang, P., Chu, C.-C., Xi, T. (2017). A novel biodegradable and biologically functional arginine-based poly(ester urea urethane) coating for Mg–Zn–Y–Nd alloy: Enhancement in corrosion resistance and biocompatibility, *J. Mater. Chem. B*, 5(9), pp. 1787–802, Doi: 10.1039/C6TB03147A.

[368] Liu, J., Wang, P., Chu, C.-C., Xi, T. (2017). Arginine-leucine based poly (ester urea urethane) coating for Mg-Zn-Y-Nd alloy in cardiovascular stent applications, *Colloids Surf. B Biointerfaces*, 159, pp. 78–88, Doi: 10.1016/J.COLSURFB.2017.07.031.

[369] Wang, P., Xiong, P., Liu, J., Gao, S., Xi, T., Cheng, Y. (2018). A silk-based coating containing GREDVY peptide and heparin on Mg–Zn–Y–Nd alloy: Improved corrosion resistance, hemocompatibility and endothelialization, *J. Mater. Chem. B*, 6 (6), pp. 966–78, Doi: 10.1039/C7TB02784B.

[370] Song, X., Chang, L., Wang, J., Zhu, S., Wang, L., Feng, K., Luo, Y., Guan, S. (2018). Investigation on the in vitro cytocompatibility of Mg-Zn-Y-Nd-Zr alloys as degradable orthopaedic implant materials, *J. Mater. Sci. Mater. Med.*, 29(4), p. 44, Doi: 10.1007/s10856-018-6050-8.

[371] Kondori, B., Mahmudi, R. (2010). Effect of Ca additions on the microstructure, thermal stability and mechanical properties of a cast AM60 magnesium alloy, *Mater. Sci. Eng. A*, 527(7–8), pp. 2014–21, Doi: 10.1016/J.MSEA.2009.11.043.

[372] Xu, S.W., Matsumoto, N., Yamamoto, K., Kamado, S., Honma, T., Kojima, Y. (2009). High temperature tensile properties of as-cast Mg–Al–Ca alloys, *Mater. Sci. Eng. A*, 509(1–2), pp. 105–10, Doi: 10.1016/J.MSEA.2009.02.024.

[373] Yang, J., Peng, J., Nyberg, E.A., Pan, F. (2016). Effect of Ca addition on the corrosion behavior of Mg–Al–Mn alloy, *Appl. Surf. Sci.*, 369, pp. 92–100, Doi: 10.1016/J.APSUSC.2016.01.283.

[374] Zhang, L., Deng, K., Nie, K., Xu, F., Su, K., Liang, W. (2015). Microstructures and mechanical properties of Mg–Al–Ca alloys affected by Ca/Al ratio, *Mater. Sci. Eng. A*, 636, pp. 279–88, Doi: 10.1016/J.MSEA.2015.03.100.

[375] Xin, Y., Hu, T., Chu, P.K. (2011). In vitro studies of biomedical magnesium alloys in a simulated physiological environment: A review, *Acta Biomater.*, 7(4), pp. 1452–9, Doi: 10.1016/J.ACTBIO.2010.12.004.

[376] Gil-Santos, A., Szakacs, G., Moelans, N., Hort, N., Van der Biest, O. (2017). Microstructure and mechanical characterization of cast Mg-Ca-Si alloys, *J. Alloys Compd.*, 694, pp. 767–76, Doi: 10.1016/J.JALLCOM.2016.10.059.

[377] U&I Corporation. The efficacy and safety of magnesium alloy screw as a novel bioabsorbable material in patients due to hand fractures – Tabular view – Clinical

Trials.gov. Available at: https://clinicaltrials.gov/ct2/show/record/NCT02456415. [accessed October 11, 2018].

[378] Li, Y., Hodgson, P.D., Wen, C. (2011). The effects of calcium and yttrium additions on the microstructure, mechanical properties and biocompatibility of biodegradable magnesium alloys, *J. Mater. Sci.*, 46(2), pp. 365–71, Doi: 10.1007/s10853-010-4843-3.

[379] Miyazaki, T., Kaneko, J., Sugamata, M. (1994). Structures and properties of rapidly solidified Mg Ca based alloys, *Mater. Sci. Eng. A*, 181–182, pp. 1410–4, Doi: 10.1016/0921-5093(94)90874-5.

[380] Du, H., Wei, Z., Liu, X., Zhang, E. (2011). Effects of Zn on the microstructure, mechanical property and bio-corrosion property of Mg–3Ca alloys for biomedical application, *Mater. Chem. Phys.*, 125(3), pp. 568–75, Doi: 10.1016/J.MATCHEMPHYS.2010.10.015.

[381] Xu, S.W., Oh-ishi, K., Sunohara, H., Kamado, S. (2012). Extruded Mg–Zn–Ca–Mn alloys with low yield anisotropy, *Mater. Sci. Eng. A*, 558, pp. 356–65, Doi: 10.1016/J.MSEA.2012.08.012.

[382] Xu, S.W., Oh-ishi, K., Kamado, S., Uchida, F., Homma, T., Hono, K. (2011). High-strength extruded Mg–Al–Ca–Mn alloy, *Scr. Mater.*, 65(3), pp. 269–72, Doi: 10.1016/J.SCRIPTAMAT.2011.04.026.

[383] Yin, D.L., Wang, J.T., Liu, J.Q., Zhao, X. (2009). On tension–Compression yield asymmetry in an extruded Mg–3Al–1Zn alloy, *J. Alloys Compd.*, 478(1–2), pp. 789–95, Doi: 10.1016/J.JALLCOM.2008.12.033.

[384] Pan, H., Qin, G., Ren, Y., Wang, L., Sun, S., Meng, X. (2015). Achieving high strength in indirectly-extruded binary Mg–Ca alloy containing Guinier–Preston zones, *J. Alloys Compd.*, 630, pp. 272–6, Doi: 10.1016/J.JALLCOM.2015.01.068.

[385] Okamoto, H., Schlesinger, M.E., Mueller, E.M. (2016). *ASM handbook, volume 3, alloy phase diagrams*, Cleveland, OH, ASM International.

[386] Kim, J.H., Lee, J.Y., Lee, K.M., Park, S.W., Lim, H.P., Park, C., Bae, J.C., Yun, K.D. (2016). Biological evaluation of anodized biodegradable magnesium-calcium alloys, *Int. Sci. Congr. Exhib.*, APMAS2015, p. 129, Doi: 10.12693/APhysPolA.129.728.

[387] Harandi, S.E., Mirshahi, M., Koleini, S., Idris, M.H., Jafari, H., Kadir, M.R.A. (2012). Effect of calcium content on the microstructure, hardness and in-vitro corrosion behavior of biodegradable Mg-Ca binary alloy, *Mater. Res.*, 16(1), pp. 11–18, Doi: 10.1590/S1516-14392012005000151.

[388] Hartwig, A. (2001). Role of magnesium in genomic stability, *Mutat. Res. Mol. Mech. Mutagen.*, 475(1–2), pp. 113–21, Doi: 10.1016/S0027-5107(01)00074-4.

[389] Okuma, T. (2001). Magnesium and bone strength, *Nutrition*, 17(7–8), pp. 679–80, Doi: 10.1016/S0899-9007(01)00551-2.

[390] Saris, N.E., Mervaala, E., Karppanen, H., Khawaja, J.A., Lewenstam, A. (2000). Magnesium. An update on physiological, clinical and analytical aspects, *Clin. Chim. Acta.*, 294(1–2), pp. 1–26.

[391] Vormann, J. (2003). Magnesium: Nutrition and metabolism, *Mol. Aspects Med.*, 24(1–3), pp. 27–37, Doi: 10.1016/S0098-2997(02)00089-4.

[392] Witte, F., Ulrich, H., Rudert, M., Willbold, E. (2007). Biodegradable magnesium scaffolds: Part 1: Appropriate inflammatory response, *J. Biomed. Mater. Res. Part A*, 81A(3), pp. 748–56, Doi: 10.1002/jbm.a.31170.

[393] Witte, F., Ulrich, H., Palm, C., Willbold, E. (2007). Biodegradable magnesium scaffolds: Part II: Peri-implant bone remodeling, *J. Biomed. Mater. Res. Part A*, 81A(3), pp. 757–65, Doi: 10.1002/jbm.a.31293.

[394] Nayab, S.N., Jones, F.H., Olsen, I. (2007). Modulation of the human bone cell cycle by calcium ion-implantation of titanium, *Biomaterials*, 28(1), pp. 38–44, Doi: 10.1016/J.BIOMATERIALS.2006.08.032.

[395] Gu, X.N., Li, X.L., Zhou, W.R., Cheng, Y., Zheng, Y.F. (2010). Microstructure, biocorrosion and cytotoxicity evaluations of rapid solidified Mg–3Ca alloy ribbons as a biodegradable material, *Biomed. Mater.*, 5(3), p. 035013, Doi: 10.1088/1748-6041/5/3/035013.

[396] Liu, Y., Liu, D., Zhao, Y., Chen, M. (2015). Corrosion degradation behavior of Mg–Ca alloy with high Ca content in SBF, *Trans. Nonferrous Met. Soc. China*, 25(10), pp. 3339–47, Doi: 10.1016/S1003-6326(15)63968-1.

[397] Cho, S.Y., Chae, S.-W., Choi, K.W., Seok, H.K., Kim, Y.C., Jung, J.Y., Yang, S.J., Kwon, G.J., Kim, J.T., Assad, M. (2013). Biocompatibility and strength retention of biodegradable Mg-Ca-Zn alloy bone implants, *J. Biomed. Mater. Res. Part B Appl. Biomater.*, 101B(2), pp. 201–12, Doi: 10.1002/jbm.b.32813.

[398] Berglund, I.S., Jacobs, B.Y., Allen, K.D., Kim, S.E., Pozzi, A., Allen, J.B., Manuel, M.V. (2016). Peri-implant tissue response and biodegradation performance of a Mg-1.0Ca-0.5Sr alloy in rat tibia, *Mater. Sci. Eng. C*, 62, pp. 79–85, Doi: 10.1016/j.msec.2015.12.002.

[399] He, G., Wu, Y., Zhang, Y., Zhu, Y., Liu, Y., Li, N., Li, M., Zheng, G., He, B., Yin, Q., Zheng, Y., Mao, C. (2015). Addition of Zn to the ternary Mg-Ca-Sr alloys significantly improves their antibacterial property, *J. Mater. Chem. B*, 3(32), pp. 6676–89, Doi: 10.1039/C5TB01319D.

[400] Wang, Q., Chen, J., Zhao, Z., He, S. (2010). Microstructure and super high strength of cast Mg-8.5Gd-2.3Y-1.8Ag-0.4Zr alloy, *Mater. Sci. Eng. A*, 528(1), pp. 323–8, Doi: 10.1016/J.MSEA.2010.09.004.

[401] Yamada, K., Hoshikawa, H., Maki, S., Ozaki, T., Kuroki, Y., Kamado, S., Kojima, Y. (2009). Enhanced age-hardening and formation of plate precipitates in Mg–Gd–Ag alloys, *Scr. Mater.*, 61(6), pp. 636–9, Doi: 10.1016/J.SCRIPTAMAT.2009.05.044.

[402] Zheng, K.Y., Dong, J., Zeng, X.Q., Ding, W.J. (2008). Precipitation and its effect on the mechanical properties of a cast Mg–Gd–Nd–Zr alloy, *Mater. Sci. Eng. A*, 489(1–2), pp. 44–54, Doi: 10.1016/J.MSEA.2007.11.080.

[403] Gao, L., Chen, R.S., Han, E.H. (2009). Effects of rare-earth elements Gd and Y on the solid solution strengthening of Mg alloys, *J. Alloys Compd.*, 481(1–2), pp. 379–84, Doi: 10.1016/J.JALLCOM.2009.02.131.

[404] Nie, J.F., Muddle, B.C. (1999). Precipitation in magnesium alloy WE54 during isothermal ageing at 250°C, *Scr. Mater.*, 40(10), pp. 1089–94, Doi: 10.1016/S1359-6462(99)00084-6.

[405] Riontino, G., Massazza, M., Lussana, D., Mengucci, P., Barucca, G., Ferragut, R. (2008). A novel thermal treatment on a Mg–4.2Y–2.3Nd–0.6Zr (WE43) alloy, *Mater. Sci. Eng. A*, 494(1–2), pp. 445–8, Doi: 10.1016/J.MSEA.2008.04.043.

[406] Wu, W., Petrini, L., Gastaldi, D., Villa, T., Vedani, M., Lesma, E., Previtali, B., Migliavacca, F. (2010). Finite element shape optimization for biodegradable magnesium alloy stents, *Ann. Biomed. Eng.*, 38(9), pp. 2829–40, Doi: 10.1007/s10439-010-0057-8.

[407] Campos, C.M., Muramatsu, T., Iqbal, J., Zhang, Y.-J., Onuma, Y., Garcia-Garcia, H.M., Haude, M., Lemos, P.A., Warnack, B., Serruys, P.W. (2013). Bioresorbable drug-eluting magnesium-alloy scaffold for treatment of coronary artery disease, *Int. J. Mol. Sci.*, 14(12), pp. 24492–500, Doi: 10.3390/ijms141224492.

[408] Biber, R., Pauser, J., Geßlein, M., Bail, H.J. (2016). Magnesium-based absorbable metal screws for intra-articular fracture fixation, *Case Rep. Orthop.*, 2016, p. 9673174, Doi: 10.1155/2016/9673174.

[409] Okamoto, H. (1992). Mg-y (magnesium-yttrium), *J. Phase Equilibria*, 13(1), pp. 105–6, Doi: 10.1007/BF02645395.

[410] Okamoto, H. (1993). Gd-Mg (gadolinium-magnesium), *J. Phase Equilibria*, 14(4), pp. 534–5, Doi: 10.1007/BF02671981.

[411] Kang, Y.-H., Yan, H., Chen, R.-S. (2015). Effects of heat treatment on the precipitates and mechanical properties of sand-cast Mg–4Y–2.3Nd–1Gd–0.6Zr magnesium alloy, *Mater. Sci. Eng. A*, 645, pp. 361–8, Doi: 10.1016/J. MSEA.2015.08.041.

[412] Zhu, Y.M., Morton, A.J., Nie, J.F. (2008). Improvement in the age-hardening response of Mg–Y–Zn alloys by Ag additions, *Scr. Mater.*, 58(7), pp. 525–8, Doi: 10.1016/J.SCRIPTAMAT.2007.11.003.

[413] Zhang, Y., Rong, W., Wu, Y., Peng, L., Nie, J.-F., Birbilis, N. (2018). A comparative study of the role of Ag in microstructures and mechanical properties of Mg-Gd and Mg-Y alloys, *Mater. Sci. Eng. A*, 731, pp. 609–22, Doi: 10.1016/J.MSEA.2018.06.084.

[414] Pan, F., Yang, M., Chen, X. (2016). A review on casting magnesium alloys: Modification of commercial alloys and development of new alloys, *J. Mater. Sci. Technol.*, 32(12), pp. 1211–21, Doi: 10.1016/J.JMST.2016.07.001.

[415] Panigrahi, S.K., Yuan, W., Mishra, R.S., DeLorme, R., Davis, B., Howell, R.A., Cho, K. (2011). A study on the combined effect of forging and aging in Mg–Y–RE alloy, *Mater. Sci. Eng. A*, 530, pp. 28–35, Doi: 10.1016/J.MSEA.2011.08.065.

[416] Barnett, M.R., Sullivan, A., Stanford, N., Ross, N., Beer, A. (2010). Texture selection mechanisms in uniaxially extruded magnesium alloys, *Scr. Mater.*, 63(7), pp. 721–4, Doi: 10.1016/J.SCRIPTAMAT.2010.01.018.

[417] Wu, B.L., Zhao, Y.H., Du, X.H., Zhang, Y.D., Wagner, F., Esling, C. (2010). Ductility enhancement of extruded magnesium via yttrium addition, *Mater. Sci. Eng. A*, 527(16–17), pp. 4334–40, Doi: 10.1016/J.MSEA.2010.03.054.

[418] Cottam, R., Robson, J., Lorimer, G., Davis, B. (2008). Dynamic recrystallization of Mg and Mg–Y alloys: Crystallographic texture development, *Mater. Sci. Eng. A*, 485(1–2), pp. 375–82, Doi: 10.1016/J.MSEA.2007.08.016.

[419] Hort, N., Huang, Y., Fechner, D., Störmer, M., Blawert, C., Witte, F., Vogt, C., Drücker, H., Willumeit, R., Kainer, K.U., Feyerabend, F. (2010). Magnesium alloys as implant materials – Principles of property design for Mg–RE alloys, *Acta Biomater.*, 6(5), pp. 1714–25, Doi: 10.1016/J.ACTBIO.2009.09.010.

[420] Chang, J., Guo, X., He, S., Fu, P., Peng, L., Ding, W. (2008). Investigation of the corrosion for Mg–xGd–3Y–0.4Zr (x = 6, 8, 10, 12 wt%) alloys in a peak-aged condition, *Corros. Sci.*, 50(1), pp. 166–77, Doi: 10.1016/J.CORSCI.2007.06.003.

[421] Takenaka, T., Ono, T., Narazaki, Y., Naka, Y., Kawakami, M. (2007). Improvement of corrosion resistance of magnesium metal by rare earth elements, *Electrochim. Acta*, 53(1), pp. 117–21, Doi: 10.1016/J.ELECTACTA.2007.03.027.

[422] Birbilis, N., Easton, M.A., Sudholz, A.D., Zhu, S.M., Gibson, M.A. (2009). On the corrosion of binary magnesium-rare earth alloys, *Corros. Sci.*, 51(3), pp. 683–9, Doi: 10.1016/J.CORSCI.2008.12.012.

[423] Liu, M., Schmutz, P., Uggowitzer, P.J., Song, G., Atrens, A. (2010). The influence of yttrium (Y) on the corrosion of Mg–Y binary alloys, *Corros. Sci.*, 52(11), pp. 3687–701, Doi: 10.1016/J.CORSCI.2010.07.019.

[424] Chou, D.-T., Hong, D., Saha, P., Ferrero, J., Lee, B., Tan, Z., Dong, Z., Kumta, P.N. (2013). In vitro and in vivo corrosion, cytocompatibility and mechanical properties of biodegradable Mg–Y–Ca–Zr alloys as implant materials, *Acta Biomater.*, 9(10), pp. 8518–33, Doi: 10.1016/J.ACTBIO.2013.06.025.

[425] Kubásek, J., Vojtěch, D. (2013). Structural and corrosion characterization of biodegradable Mg–RE (RE=Gd, Y, Nd) alloys, *Trans. Nonferrous Met. Soc. China*, 23(5), pp. 1215–25, Doi: 10.1016/S1003-6326(13)62586-8.

[426] Schroeder, H.A., Mitchener, M. (1971). Scandium, chromium(VI), gallium, yttrium, rhodium, palladium, indium in mice: Effects on growth and life span, *J. Nutr.*, 101(10), pp. 1431–7.

[427] Naujokat, H., Gülses, A., Wiltfang, J., Açil, Y. (2017). Effects of degradable osteosynthesis plates of MgYREZr alloy on cell function of human osteoblasts, fibroblasts and osteosarcoma cells, *J. Mater. Sci. Mater. Med.*, 28(8), p. 126, Doi: 10.1007/s10856-017-5938-z.

[428] Castellani, C., Lindtner, R.A., Hausbrandt, P., Tschegg, E., Stanzl-Tschegg, S.E., Zanoni, G., Beck, S., Weinberg, A.-M. (2011). Bone–Implant interface strength and osseointegration: Biodegradable magnesium alloy versus standard titanium control, *Acta Biomater.*, 7(1), pp. 432–40, Doi: 10.1016/J.ACTBIO.2010.08.020.

[429] Ye, C.-H., Xi, T.-F., Zheng, Y.-F., Wang, S.-Q., Li, Y.-D. (2013). In vitro corrosion and biocompatibility of phosphating modified WE43 magnesium alloy, *Trans. Nonferrous Met. Soc. China*, 23, pp. 996–1001, Doi: 10.1016/S1003-6326(13)62558-3.

[430] Ye, C.H., Zheng, Y.F., Wang, S.Q., Xi, T.F., Li, Y.D. (2012). In vitro corrosion and biocompatibility study of phytic acid modified WE43 magnesium alloy, *Appl. Surf. Sci.*, 258(8), pp. 3420–7, Doi: 10.1016/J.APSUSC.2011.11.087.

[431] Zhen, Z., Xi, T.F., Zheng, Y.F. (2015). Surface modification by natural biopolymer coatings on magnesium alloys for biomedical applications, *Surf. Modif. Magnes. Its Alloy. Biomed. Appl.*, pp. 301–33, Doi: 10.1016/B978-1-78242-078-1.00011-6.

[432] Dobatkin, S.V., Lukyanova, E.A., Martynenko, N.S., Anisimova, N.Y., Kiselevskiy, M.V., Gorshenkov, M.V., Yurchenko, N.Y., Raab, G.I., Yusupov, V.S., Birbilis, N., Salishchev, G.A., Estrin, Y.Z. (2017). Strength, corrosion resistance, and biocompatibility of ultrafine-grained Mg alloys after different modes of severe plastic deformation, *IOP Conf. Ser. Mater. Sci. Eng.*, 194(1), p. 012004, Doi: 10.1088/1757-899X/194/1/012004.

[433] Smola, B., Joska, L., Březina, V., Stulíková, I., Hnilica, F. (2012). Microstructure, corrosion resistance and cytocompatibility of Mg–5Y–4Rare Earth–0.5Zr (WE54) alloy, *Mater. Sci. Eng. C*, 32(4), pp. 659–64, Doi: 10.1016/J.MSEC.2012.01.003.

4 Tackling the Challenges

4.1 INTRODUCTION

In Chapter 3, the main alloys studied have been reviewed from a biological perspective, aiming to provide the readers with useful guidelines for the choice of the material that best fits their applications according to the state of the art. Many alloys have been studied in recent years, but a thorough and concise presentation of the different materials' pros and cons and their comparison is missing. In this chapter, the authors provide a radar chart that compares the corrosion, mechanical and biological performances of the studied alloys, which will give the readers a better oversight of the material that best fits their requirements.

4.2 RADAR CHART: AN EASY TOOL TO COMPARE CORROSION, MECHANICAL AND BIOLOGICAL PERFORMANCES

In Figure 4.1, the main alloying families (high-pure magnesium, aluminum-based alloys, zinc-based alloys, calcium-based alloys and rare earth (RE)-based alloys) reviewed herein are gathered in a radar chart that allows a clear and fast understanding of their performances with respect to the mechanical properties, corrosion resistance and biocompatibility. In this way, it is possible for the reader to understand the best material for his/her applications.

It can be noted that the overall properties of the Zn-based alloys exceed those of the other Mg alloys. Ultra-high-purity Mg alloys are effective in improving the corrosion resistance. However, purification only, in particular that of pure magnesium, is not enough to produce potential biomedical materials, as they also possess low mechanical properties. Here, thermo-mechanical treatments could be the solution (see below and Section 3.2.1, Chapter 3). Calcium-based alloys have been considered great candidates for biomedical implants due to the high biocompatibility of calcium, but their mechanical properties are lower than those of Zn-based and RE-based alloys. In addition, calcium-based alloys are not so effective with respect to the corrosion resistance, as they can easily form second phases because of the very low solubility of calcium. The main drawback of Al-based Mg alloys is their biocompatibility: long-term effects of exposure to aluminum reveals aluminum to be toxic, affecting the reproductive ability [1], inducing dementia [2] and leading to Alzheimer's disease [3,4]. Dealing with RE-based alloys, their mechanical properties and corrosion resistance are comparable to those of Zn-based alloys, but their biocompatibility is still debatable since some RE elements might induce latent toxic and harmful effects on the human body during degradation [5,6]. Among Zn-based Mg alloys, Mg–Zn–Zr alloys are very interesting due to the great grain refinement effect of zirconia, but high attention must be paid not to exceed the low solubility content of Zr, so as to avoid the

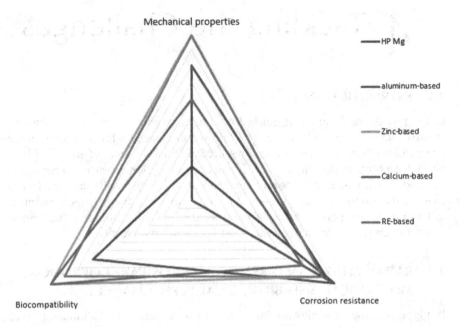

FIGURE 4.1 Radar chart of the main alloying families with respect to the mechanical properties, corrosion resistance and biocompatibility.

formation of second phases. In addition, Mg–Zn–RE alloys have been greatly considered for cardiovascular applications because of the large elongation that they provide as a consequence of the randomized texture obtained after dynamic recrystallization.

Once the best suitable material is determined, its properties have to be optimized. However, as described in Chapter 2, it is evidenced that some methodologies adopted to improve the mechanical performances negatively affect the corrosion performances and vice versa. Increasing the concentration of alloying elements leads to increased strength, while the formation of second phases leads to an increment in the corrosion rate. Considering, for example, Mg–Zn–Zr alloys, Yu et al. [7] reported that increasing the Zn concentration from 2 wt.% to 6 wt.% leads to an increment in ultimate tensile strength from 308 to 360 MPa. However, increasing the Zn content up to 6 wt.% corresponds to exceeding its solubility limit, determining an increment in corrosion rate of more than one order of magnitude [8]. For each material, it is thus fundamental to define the optimal concentration of the alloying elements, considering a trade-off between these two properties. This has always been a challenging task; however, from the literature analysis conducted, some guidelines can be provided. To minimize the corrosion rate, the alloying elements have to be added up to an amount lower than their solubility limit. However, this strategy would exclude the contribution of the precipitation hardening from the strengthening mechanisms. Nevertheless, the contribution of the precipitation hardening can be easily substituted by further leveraging on the grain refinement strengthening, achievable through severe plastic deformation (SPD) techniques. Among the different strengthening mechanisms, grain refinement was found to be the key factor affecting the mechanical response of magnesium and its alloys [9]. Furthermore, decreasing the grain size also has a strong effect on reducing the corrosion rate. However, in some cases, ultra-fine-grained Mg alloys have been reported to be characterized by a corrosion rate higher than their counterparts characterized by slightly coarser grains. For example, Seong and Kim [10] reported that the corrosion rate of the extruded and HRDSR-processed Mg–2Ca alloy was three times higher than that of their counterparts extruded alone. This is due to the increased dislocation densities during SPD by HRDSR. In fact, the dislocations in the matrices of Mg alloys have been suggested to play a negative role in corrosion resistance because the regions with a higher concentration of dislocations are anodic to the regions with lower concentrations [11]; thus, more highly deformed regions will corrode faster. However, carrying out annealing treatments has revealed to be functional in the improvement of the corrosion behavior: the corrosion rate of the annealed extruded and HRDSR-processed Mg–2Ca alloys was reduced by an order of magnitude due to the reduction of the dislocation density. After the optimization of its parameters [12], the annealing process slightly increases the grain size, but this procedure allows to obtain the best trade-off between mechanical and corrosion properties.

REFERENCES

[1] Domingo, J.L. (1995). Reproductive and developmental toxicity of aluminum: A review, *Neurotoxicol. Teratol.*, 17(4), pp. 515–21, Doi: 10.1016/0892-0362(95)00002-9.

[2] Venugopal, B., Luckey, T.D. (1978). *Metal toxicity in mammals. Volume 2. Chemical toxicity of metals and metalloids*, New York, Plenum Press.

[3] Flaten, T.P. (2001). Aluminium as a risk factor in Alzheimer's disease, with emphasis on drinking water, *Brain Res. Bull.*, 55(2), pp. 187–96.

[4] El-Rahman, S.S.A. (2003). Neuropathology of aluminum toxicity in rats (glutamate and GABA impairment), *Pharmacol. Res.*, 47(3), pp. 189–94.

[5] Nakamura, Y., Tsumura, Y., Tonogai, Y., Shibata, T., Ito, Y. (1997). Differences in behavior among the chlorides of seven rare earth elements administered intravenously to rats, *Fundam. Appl. Toxicol.*, 37(2), pp. 106–16, Doi: 10.1006/FAAT.1997.2322.

[6] Xu, Z., Smith, C., Chen, S., Sankar, J. (2011). Development and microstructural characterizations of Mg–Zn–Ca alloys for biomedical applications, *Mater. Sci. Eng. B*, 176(20), pp. 1660–5, Doi: 10.1016/J.MSEB.2011.06.008.

[7] Yu, Z.H., Yan, H.G., Chen, J.H., Wu, Y.Z. (2010). Effect of Zn content on the microstructures and mechanical properties of laser beam-welded ZK series magnesium alloys, *J. Mater. Sci.*, 45(14), pp. 3797–803, Doi: 10.1007/s10853-010-4434-3.

[8] Huan, Z.G., Leeflang, M.A., Zhou, J., Fratila-Apachitei, L.E., Duszczyk, J. (2010). In vitro degradation behavior and cytocompatibility of Mg-Zn-Zr alloys, *J. Mater. Sci. Mater. Med.*, 21(9), pp. 2623–35, Doi: 10.1007/s10856-010-4111-8.

[9]   Markushev, M.V., Nugmanov, D.R., Sitdikov, O., Vinogradov, A. (2018). Struc-
      ture, texture and strength of Mg-5.8Zn-0.65Zr alloy after hot-to-warm multi-step
      isothermal forging and isothermal rolling to large strains, *Mater. Sci. Eng. A*,
      709, pp. 330–8, Doi: 10.1016/J.MSEA.2017.10.008.

[10] Seong, J.W., Kim, W.J. (2015). Development of biodegradable Mg–Ca alloy sheets
      with enhanced strength and corrosion properties through the refinement and uni-
      form dispersion of the Mg2Ca phase by high-ratio differential speed rolling, *Acta
      Biomater.*, 11, pp. 531–42, Doi: 10.1016/J.ACTBIO.2014.09.029.

[11] Andrei, M., Eliezer, A., Bonora, P.L., Gutman, E.M. (2002). DC and AC polarisa-
      tion study on magnesium alloys influence of the mechanical deformation, *Mater.
      Corros.*, 53(7), pp. 455–61, Doi: 10.1002/1521-4176(200207)53:7<455::AID-
      MACO455>3.0.CO;2-4.

[12] Chen, H., Yu, H., Kang, S.B., Cho, J.H., Min, G. (2010). Optimization of annealing
      treatment parameters in a twin roll cast and warm rolled ZK60 alloy sheet, *Mater.
      Sci. Eng. A*, 527(4–5), pp. 1236–42, Doi: 10.1016/J.MSEA.2009.09.057.

5 Outlook

The applicability of magnesium and its alloys as materials for temporary implants is hampered by several issues, such as the low mechanical properties and high corrosion rate, and the efforts in this field are continuously increasing. Some researchers are focusing on the application of biocompatible surface coatings, such as TiO_2, ZrO_2 and HfO_2, obtained through conversion coatings, sol–gel coatings, chemical vapor and physical vapor deposition and plasma electrolytic oxidation coatings. Coating procedures are reported to highly increase the corrosion resistance of the alloys. Bakhsheshi-Rad et al. [1], in fact, reported the corrosion current density of physical vapor deposition-coated Mg-3Zn-0.8Ca to be five times lower than that of their bare counterparts. However, coatings have to be uniformly deposited to provide a barrier to the corrosive attacks of body fluids. Furthermore, coatings usually suffer from limited stability under cyclic loading in physiological conditions. These factors represent the main drawbacks that hamper the applicability of coatings. Instead, the authors believe that future studies will delve deep into advanced plastic deformation techniques working directly on the Mg material and aiming to optimize its mechanical and corrosion properties. In addition, researchers will intensify efforts on investigating the corrosion-assisted cracking phenomena, that is, stress corrosion cracking and corrosion fatigue, which, so far, have been underexplored. Hence, the optimization of the severe plastic deformation techniques must be considered in a wider aspect, not only aiming to optimize mechanical properties and corrosion resistance as two separate entities but as two interconnected ones, because the stress corrosion cracking susceptibility is related to both mechanical and corrosion properties. In addition, because of the harmful effects of impurities and second phases on the corrosion resistance, the authors envision that another key research line will be based on ultra-high-pure magnesium and ultra-high-pure alloys, with a composition avoiding the formation of second phases entirely.

REFERENCE

[1] Bakhsheshi-Rad, H.R., Hamzah, E., Dias, G.J., Saud, S.N., Yaghoubidoust, F., Hadisi, Z. (2017). Fabrication and characterisation of novel ZnO/MWCNT duplex coating deposited on Mg alloy by PVD coupled with dip-coating techniques, *J. Alloys Compd.*, 728, pp. 159–68, Doi: 10.1016/J.JALLCOM.2017.08.161.

Appendix A: Corrosion

A.1 CORROSION PROCESS

When magnesium is exposed to an aqueous environment, it corrodes according to the following reaction, developing a dull gray layer composed of magnesium hydroxide ($Mg(OH)_2$) [1,2].

$$Mg + 2H_2O \rightarrow Mg(OH)_2 + H_2 \qquad (A.1)$$

The corrosion reaction can be divided into anodic and cathodic partial reactions, (A.2) and (A.3), respectively [3].

$$Mg \rightarrow Mg^{2+} + 2e^- \qquad (A.2)$$

$$2H_2O + 2e^- \rightarrow H_2 + 2(OH)^- \qquad (A.3)$$

The $Mg(OH)_2$ film formed on the surface of magnesium and its alloys prevents further corrosion in water since it is slightly soluble there (0.01 mm/year [4]). However, if the corrosive medium contains any chlorides with concentration of above 30 mmol/L [5], then $Mg(OH)_2$ will be converted to magnesium chloride ($MgCl_2$) according to reference [6]:

$$Mg(OH)_2 + 2Cl^- \rightarrow MgCl_2 + 2(OH)^- \qquad (A.4)$$

Magnesium chloride is highly soluble in aqueous solution, determining further corrosion development of magnesium and its alloys in the human body, where the chloride content is about 150 mmol/L [7–9]. Besides the need of being characterized by a low corrosion rate to ensure sufficient residual strength until healing has occurred and to avoid the necrosis of tissues due to a high hydrogen evolution rate, uniform corrosion is also highly required. Nonuniform corrosion is in fact more dangerous for the devices subjected to mechanical loads, as nonuniform corrosion site is preferential for the onset of failure under quasi-static loading (stress corrosion cracking phenomenon) and fatigue (corrosion fatigue phenomenon). However, uniform corrosion is not the main corrosion mode. Kirkland et al. [10] reported that 29 of 31 magnesium alloys suffer from localized corrosion. In the following section, the main nonuniform corrosion modes that affect magnesium alloys are gathered.

A.2 NON-UNIFORM CORROSION MODES

A.2.1 Intergranular Corrosion

Intergranular corrosion is a particular type of galvanic corrosion. Galvanic corrosion takes place between two dissimilar metals when they come in contact in the presence of an electrolyte, which is the body fluid. The less noble metal becomes anodic and thus corrodes, and, magnesium being the most reactive metal in the electrochemical series, it will always be the anode in any corrosion reactions [11]. Second phases, inclusions and impurities deposited on the grain boundary region during solidification are more cathodic with respect to the magnesium matrix, leading to the occurrence of intergranular corrosion (Figure A.1).

A.2.2 Pitting Corrosion

This type of corrosion results from the rapid corrosion of small localized areas, which damages the protective surface oxide layer. The damage of the magnesium hydroxide layer can be due to the presence of chlorides or due to the effect of the applied loads. After initial nucleation at the surface, the presence of impurities in the alloy microstructures leads to further growing of the pits [12,13]. Rarely, localized corrosion occurs in the absence of second phases, precipitates or impurities. Furthermore, as diluting the pit contents is impossible because of its small mouth, the autocatalytic growth of the pit is aggravated. In addition, localized production of the positive metal ions in the pit gives a local excess of positive charge, which attracts the negative chloride ions to produce charge neutrality, further breaking the protective layer and accelerating the pits growth (Figure A.2). This form of corrosion is very dangerous since the pits on the surface are very small and their detection is difficult because they are often hidden by corrosion products.

A.2.3 Tribocorrosion

Tribocorrosion is a material degradation process caused by the combined effect of corrosion and wear. Because of the repeated relative surface motions, as

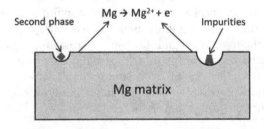

FIGURE A.1 Schematic representation of intergranular corrosion.

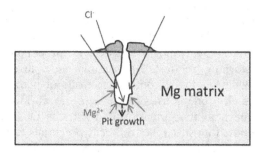

FIGURE A.2 Schematic representation of pitting corrosion.

induced, for example, by the normal everyday activities experienced by the human bodies, the magnesium hydroxide and part of the fresh metal are normally damaged and removed. This leads to wear and debris between the surfaces that have further detrimental effects since they act as an abrasive agent. Tribocorrosion is common in load-bearing applications [14,15], and the corrosion rate is dependent on the applied load, the relative motion, the microstructure of the alloy and the chemistry of the corrosive environment [16].

A.3 CORROSION MEASUREMENT TECHNIQUES

There are several techniques used to evaluate the corrosion of a material, but here we will describe only the procedures considered in the prosecution of the work since they are the most common, for example, potentiodynamic polarization test, mass loss experiments and hydrogen evolution tests. The latter two are used in the determination of the corrosion rate. The former can also be linked to the corrosion rate of the material, but it cannot take into account the evolution of the corrosion behavior over the time of exposure to the corrosive environment. Further techniques can be found in ref. [17].

A.3.1 POTENTIODYNAMIC POLARIZATION TEST

This technique is an electrochemical measurement. The corrosion process is in fact governed by two opposing electrochemical reactions: the anodic reaction and the cathodic reaction. When these two reactions are in equilibrium, the flow of electrons from each reaction is balanced, and no net electron flow (electrical current) occurs. The potential at which the anodic and cathodic reactions are kept in balance (E_{corr}) is called open-circuit potential (OCP). The value of either the anodic or cathodic current at the OCP is called corrosion current, I_{corr}. Unfortunately, I_{corr} cannot be measured directly, but it can be estimated using the Tafel extrapolation technique [18]. In addition, from the potentiodynamic polarization curves in the anodic region, it is possible to evaluate whether the material possesses a passivation behavior or not. To do so, the magnesium sample (the working electrode) with a polished surface area of few square centimeters is immersed in the physiological solution (the electrolyte). A reference electrode and a counter electrode are also immersed in the solution and all the three electrodes are connected to a device called potentiostat (sometimes a galvanostat is used). In the potentiostat, a potential is applied to the working electrode (the sample under investigation) and the current is measured through the counter electrode (a platinum wire/foil usually). The reference electrode is used to measure the working electrode potential to give a feedback to the potentiostat. First, the OCP has to be found, and after that the potentiostat applies different potentials, usually in the range of OCP ± 0.5 V, and the relative current is measured. The corrosion current, I_{corr}, or the corrosion current density, i_{corr}, is related to the corrosion rate according to the formulation reported in the ASTM G59-97. In particular, the lower the corrosion current density, the lower the corrosion rate, and it is thus clear that one of the parameters adopted to assess the corrosion behavior of magnesium and its alloys is i_{corr}. However, the corrosion rate values thus obtained have been found to be not so precise due to the negative difference effect [19,20].

A.3.2 Mass Loss Experiment

The mass loss experiment is the most simplistic method for corrosion testing of a material. The tested material is weighted before its immersion in the corrosive environment. After the immersion period, the corrosion products are removed. Dealing with magnesium, Thirumalaikumarasamy et al. [21] suggested the immersion of the specimens for one minute in a solution prepared by using 50 g chromium trioxide (CrO_3), 2.5 g silver nitrate ($AgNO_3$) and 5 g barium nitrate ($Ba(NO_3)_2$) for 250 ml distilled water as an effective method to remove the corrosion products. After this operation, the tested samples are weighted again and the mass loss W (in grams) is recorded. The corrosion rate is then measured according to the following equation:

$$CR = \frac{87,600 \times W}{A \times T \times D} \tag{A.5}$$

where CR is the corrosion rate in mm/year, A is the surface area surrounded by the corrosive fluid (cm^2), T is the time of exposure (h) and D is the density of the material (g/cm^3).

However, accurate measurement of weight loss is not so easy, as it requires complete removal of the products stuck on the specimen surface without removing any un-corroded metal, which is a difficult task. The effort to keep the un-corroded metal untouched during the corrosion product removal usually leads to incomplete removal of the corrosion products. Sometimes, while removing the corrosion products, the un-corroded areas of the metallic substrate underneath the corrosion products may also get removed. Therefore, some experimental errors might be introduced into the final weight-loss rate [22].

A.3.3   Hydrogen Evolution Test

To overcome the drawbacks of the potentiodynamic polarization tests and, above all, of the mass loss experiments, Song et al. [22] developed a new and more accurate procedure to assess the corrosion rate of magnesium and its alloys. According to Equation (A.1), the evolution of one mole of hydrogen gas corresponds to the dissolution of one mole of magnesium. Therefore, measuring the volume of hydrogen evolved is equivalent to measuring the weight-loss of magnesium dissolved, and the measured hydrogen evolution rate is equal to the weight-loss rate if both are converted into same units. An argument may exist that sometimes the falling of some small metallic particles from the magnesium surface during corrosion due to the undermining effect would cause errors to hydrogen evolution measurement. However, if the experimental set-up is designed in such a way that hydrogen evolution from the undermined particles can also be collected together with the hydrogen from the magnesium specimen, then the total amount of hydrogen collected should still be equal to the total amount of dissolved magnesium. The experimental set-up designed in ref. [22] is very simple: the magnesium specimen is put in a beaker containing the test solution, and a funnel is placed over the specimen to collect all the hydrogen from the specimen surface and from any undermined metal particles. At last, a burette is mounted over the funnel, which is filled with the test solution (Figure A.3). The hydrogen collected by the funnel goes into the burette and the evolved hydrogen can be easily measured by reading the position of the test solution level in the burette.

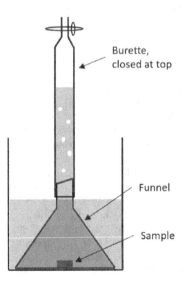

Burette,
closed at top

Funnel

Sample

FIGURE A.3   Schematic illustration of the set-up for measurement of the volume of hydrogen evolved.

The measured hydrogen volume can be easily converted into the number of gram moles by using the ideal gas laws. Simply assuming the number of moles of the evolved hydrogen is equal to the number of moles of the dissolved magnesium, the corrosion rate can be measured by substituting Equation (A.6) into Equation (A.5).

$$W = \frac{H_{\text{evolved}} \times D_{H_2}}{M_{W,H_2}} M_{W,\text{Mg}} \tag{A.6}$$

where H_{evolved} corresponds to the volume of evolved hydrogen, D_{H2} is the density of hydrogen, $M_{W,H2}$ is the molecular weight of hydrogen, and $M_{W,\text{Mg}}$ is the molecular weight of magnesium.

A.3.4 OTHER CORROSION MEASUREMENTS

Besides the procedures mentioned above, the corrosion behavior of the materials can be evaluated with other techniques. Although less accurate, the Mg^{2+} ion concentration measurements have been widely used as a comparison tool to assess the different corrosion behaviors of different alloys and to assess the effects of manufacturing strategies on these behaviors [23–36]. Researchers have now started to use electrochemical measurements not only for DC polarization curves but also for electrochemical impedance spectroscopy. In particular, they are focusing on the temporal changes of the Nyquist plot and on the modeled resistances to compare the evolution of the corrosion behavior of either different alloys or the same alloy manufactured with different processes [37–52]. Recently, researchers have started to use optical profilometry to assess the number of pits after immersion in the corrosive environment [53,54].

REFERENCES

[1] Mueller, W.-D., Lucia Nascimento, M., Lorenzo de Mele, M.F. (2010). Critical discussion of the results from different corrosion studies of Mg and Mg alloys for biomaterial applications, *Acta Biomater.*, 6(5), pp. 1749–55, Doi: 10.1016/j.actbio.2009.12.048.

[2] Chen, K., Dai, J., Zhang, X. (2015). Improvement of corrosion resistance of magnesium alloys for biomedical applications, *Corros. Rev.*, 33(3-4), pp. 101–17, Doi: 10.1515/corrrev-2015-0007.

[3] Atrens, A., Liu, M., Zainal Abidin, N.I. (2011). Corrosion mechanism applicable to biodegradable magnesium implants, *Mater. Sci. Eng. B*, 176(20), pp. 1609–36, Doi: 10.1016/j.mseb.2010.12.017.

[4] Ghali, E. (2010). *Corrosion resistance of aluminum and magnesium alloys: Understanding, performance, and testing*, Danvers, MA, John Wiley & Sons.

[5] Shaw, B.A. (2003). Corrosion resistance of magnesium alloys. In *ASM handbook*, Vol. 13A, Pennsylvania State University, pp. 692–696.

[6] Staiger, M.P., Pietak, A.M., Huadmai, J., Dias, G. (2006). Magnesium and its alloys as orthopedic biomaterials: A review, *Biomaterials*, 27, pp. 1728–34, Doi: 10.1016/j.biomaterials.2005.10.003.

[7] Witte, F., Hort, N., Vogt, C., Cohen, S., Kainer, K.U., Willumeit, R., Feyerabend, F. (2008). Degradable biomaterials based on magnesium corrosion, *Curr. Opin. Solid State Mater. Sci.*, 12(5), pp. 63–72, Doi: 10.1016/j. cossms.2009.04.001.

[8] Xu, L., Yu, G., Zhang, E., Pan, F., Yang, K. (2007). In vivo corrosion behavior of Mg-Mn-Zn alloy for bone implant application, *J. Biomed. Mater. Res. Part A*, 83A (3), pp. 703–11, Doi: 10.1002/jbm.a.31273.

[9] Witte, F., Kaese, V., Haferkamp, H., Switzer, E., Meyer-Lindenberg, A., Wirth, C.J., Windhagen, H. (2005). In vivo corrosion of four magnesium alloys and the associated bone response, *Biomaterials*, 26(17), pp. 3557–63, Doi: 10.1016/j.biomaterials.2004.09.049.

[10] Kirkland, N.T., Lespagnol, J., Birbilis, N., Staiger, M.P. (2010). A survey of bio-corrosion rates of magnesium alloys, *Corros. Sci.*, pp. 287–91, Doi: 10.1016/j. corsci.2009.09.033.

[11] Jacobs, J.J., Gilbert, J.L., Urban, R.M. (1998). Corrosion of metal orthopaedic implants, *J. Bone Joint Surg. Am.*, 80(2), pp. 268–82.

[12] Ambat, R., Aung, N.N., Zhou, W. (2000). Evaluation of microstructural effects on corrosion behaviour of AZ91D magnesium alloy, *Corros. Sci.*, 42(8), pp. 1433–55, Doi: 10.1016/S0010-938X(99)00143-2.

[13] Liu, L.J., Schlesinger, M. (2009). Corrosion of magnesium and its alloys, *Corros. Sci.*, 51, pp. 1733–7, Doi: 10.1016/j.corsci.2009.04.025.

[14] Landolt, D., Mischler, S., Stemp, M. (2001). Electrochemical methods in tribocorrosion: A critical appraisal, *Electrochim. Acta*, 46(24–25), pp. 3913–29, Doi: 10.1016/S0013-4686(01)00679-X.

[15] Yan, Y., Neville, A., Dowson, D. (2006). Biotribocorrosion – An appraisal of the time dependence of wear and corrosion interactions: II. Surface analysis, *J. Phys. D. Appl. Phys.*, 39(15), pp. 3206–12, Doi: 10.1088/0022-3727/39/15/S11.

[16] Neyman, A., Olszewski, O. (1993). Research on fretting wear dependence of hardness ratio and friction coefficient of fretted couple, *Wear*, 162–164, pp. 939–43, Doi: 10.1016/0043-1648(93)90098-7.

[17] Esmaily, M., Svensson, J.E., Fajardo, S., Birbilis, N., Frankel, G.S., Virtanen, S., Arrabal, R., Thomas, S., Johansson, L.G. (2017). Fundamentals and advances in magnesium alloy corrosion, *Prog. Mater. Sci.*, 89, pp. 92–193, Doi: 10.1016/J. PMATSCI.2017.04.011.

[18] Burstein, G.T. (2005). A hundred years of Tafel's equation: 1905–2005, *Corros. Sci.*, 47(12), pp. 2858–70, Doi: 10.1016/J.CORSCI.2005.07.002.

[19] Song, G., Atrens, A. (2003). Understanding magnesium corrosion – A framework for improved alloy performance, *Adv. Eng. Mater.*, 5(12), pp. 837–58, Doi: 10.1002/ adem.200310405.

[20] Shi, Z., Liu, M., Atrens, A. (2010). Measurement of the corrosion rate of magnesium alloys using Tafel extrapolation, *Corros. Sci.*, 52(2), pp. 579–88, Doi: 10.1016/ J.CORSCI.2009.10.016.

[21] Thirumalaikumarasamy, D., Shanmugam, K., Balasubramanian, V. (2014). Comparison of the corrosion behaviour of AZ31B magnesium alloy under immersion test and potentiodynamic polarization test in NaCl solution, *J. Magnes. Alloy.*, 2 (1), pp. 36–49, Doi: 10.1016/J.JMA.2014.01.004.

[22] Song, G., Atrens, A., StJohn, D. (2001). An hydrogen evolution method for the estimation of the corrosion rate of magnesium alloys. In *Magnesium technology*, Ed. J. N. Hryn, Hoboken, NJ, John Wiley & Sons, Inc., pp. 254–62.

[23] Feng, H., Zhang, X., Wu, G., Jin, W., Hao, Q., Wang, G., Huang, Y., Chu, P.K. (2016). Unusual anti-bacterial behavior and corrosion resistance of magnesium

alloy coated with diamond-like carbon, *RSC Adv.*, 6(18), pp. 14756–62, Doi: 10.1039/C5RA22485C.

[24] Mohajernia, S., Hejazi, S., Eslami, A., Saremi, M. (2015). Modified nanostructured hydroxyapatite coating to control the degradation of magnesium alloy AZ31 in simulated body fluid, *Surf. Coatings Technol.*, 263, pp. 54–60, Doi: 10.1016/J. SURFCOAT.2014.12.059.

[25] Wang, Z.-L., Yan, Y.-H., Wan, T., Yang, H. (2013). Poly(l -lactic acid)/hydroxyapatite/collagen composite coatings on AZ31 magnesium alloy for biomedical application, *Proc. Inst. Mech. Eng. Part H J. Eng. Med.*, 227(10), pp. 1094–103, Doi: 10.1177/0954411913493845.

[26] Qiu, X., Wan, P., Tan, L., Fan, X., Yang, K. (2014). Preliminary research on a novel bioactive silicon doped calcium phosphate coating on AZ31 magnesium alloy via electrodeposition, *Mater. Sci. Eng. C*, 36, pp. 65–76, Doi: 10.1016/J. MSEC.2013.11.041.

[27] Gray-Munro, J.E., Seguin, C., Strong, M. (2009). Influence of surface modification on the *in vitro* corrosion rate of magnesium alloy AZ31, *J. Biomed. Mater. Res. Part A*, 91A(1), pp. 221–30, Doi: 10.1002/jbm.a.32205.

[28] Razavi, M., Fathi, M., Savabi, O., Razavi, S.M., Heidari, F., Manshaei, M., Vashaee, D., Tayebi, L. (2014). In vivo study of nanostructured diopside (CaMg-Si2O6) coating on magnesium alloy as biodegradable orthopedic implants, *Appl. Surf. Sci.*, 313, pp. 60–6, Doi: 10.1016/J.APSUSC.2014.05.130.

[29] Razavi, M., Fathi, M., Savabi, O., Vashaee, D., Tayebi, L. (2015). In vitro analysis of electrophoretic deposited fluoridated hydroxyapatite coating on micro-arc oxidized AZ91 magnesium alloy for biomaterials applications, *Metall. Mater. Trans. A*, 46(3), pp. 1394–404, Doi: 10.1007/s11661-014-2694-2.

[30] Yu, W., Zhao, H., Ding, Z., Zhang, Z., Sun, B., Shen, J., Chen, S., Zhang, B., Yang, K., Liu, M., Chen, D., He, Y. (2017). In vitro and in vivo evaluation of MgF2 coated AZ31 magnesium alloy porous scaffolds for bone regeneration, *Colloid. Surf. B Biointerfaces*, 149, pp. 330–40, Doi: 10.1016/j.colsurfb.2016.10.037.

[31] Wei, Z., Tian, P., Liu, X., Zhou, B. (2015). *In vitro* degradation, hemolysis, and cytocompatibility of PEO/PLLA composite coating on biodegradable AZ31 alloy, *J. Biomed. Mater. Res. Part B Appl. Biomater.*, 103(2), pp. 342–54, Doi: 10.1002/ jbm.b.33208.

[32] Razavi, M., Fathi, M., Savabi, O., Vashaee, D., Tayebi, L. (2014). In vitro study of nanostructured diopside coating on Mg alloy orthopedic implants, *Mater. Sci. Eng. C*, 41, pp. 168–77, Doi: 10.1016/j.msec.2014.04.039.

[33] Agarwal, S., Morshed, M., Labour, M.-N., Hoey, D., Duffy, B., Curtin, J., Jaiswal, S. (2016). Enhanced corrosion protection and biocompatibility of a PLGA–silane coating on AZ31 Mg alloy for orthopaedic applications, *RSC Adv.*, 6(115), pp. 113871–83, Doi: 10.1039/C6RA24382G.

[34] Wang, S.-H., Yang, C.-W., Lee, T.-M. (2016). Evaluation of microstructural features and *in vitro* biocompatibility of hydrothermally coated fluorohydroxyapatite on AZ80 Mg alloy, *Ind. Eng. Chem. Res.*, 55(18), pp. 5207–15, Doi: 10.1021/acs. iecr.5b04583.

[35] Rojaee, R., Fathi, M., Raeissi, K. (2013). Controlling the degradation rate of AZ91 magnesium alloy via sol–gel derived nanostructured hydroxyapatite coating, *Mater. Sci. Eng. C*, 33(7), pp. 3817–25, Doi: 10.1016/J.MSEC.2013.05.014.

[36] Gu, X., Mao, Z., Ye, S.-H., Koo, Y., Yun, Y., Tiasha, T.R., Shanov, V., Wagner, W. R. (2016). Biodegradable, elastomeric coatings with controlled anti-proliferative agent release for magnesium-based cardiovascular stents, *Colloid. Surf. B Biointerfaces*, 144, pp. 170–9, Doi: 10.1016/J.COLSURFB.2016.03.086.

[37] Jang, Y., Tan, Z., Jurey, C., Collins, B., Badve, A., Dong, Z., Park, C., Kim, C.S., Sankar, J., Yun, Y. (2014). Systematic understanding of corrosion behavior of plasma electrolytic oxidation treated AZ31 magnesium alloy using a mouse model of subcutaneous implant, *Mater. Sci. Eng. C*, 45, pp. 45–55, Doi: 10.1016/j.msec.2014.08.052.

[38] Wang, X., Cai, S., Xu, G., Ye, X., Ren, M., Huang, K. (2013). Surface characteristics and corrosion resistance of sol–gel derived CaO–P2O5–SrO–Na2O bioglass–ceramic coated Mg alloy by different heat-treatment temperatures, *J. Sol-Gel Sci. Technol.*, 67(3), pp. 629–38, Doi: 10.1007/s10971-013-3122-6.

[39] Li, Y., Cai, S., Shen, S., Xu, G., Zhang, F., Wang, F. (2018). Self-healing hybrid coating of phytic acid/silane for improving the corrosion resistance of magnesium alloy, *J. Coatings Technol. Res.*, pp. 1–11, Doi: 10.1007/s11998-017-0014-7.

[40] Niu, B., Shi, P., Shanshan, E., Wei, D., Li, Q., Chen, Y. (2016). Preparation and characterization of HA sol–Gel coating on MAO coated AZ31 alloy, *Surf. Coatings Technol.*, 286, pp. 42–8, Doi: 10.1016/J.SURFCOAT.2015.11.056.

[41] Gu, X.N., Zheng, Y.F., Chen, L.J. (2009). Influence of artificial biological fluid composition on the biocorrosion of potential orthopedic Mg–Ca, AZ31, AZ91 alloys, *Biomed. Mater.*, 4(6), p. 065011, Doi: 10.1088/1748-6041/4/6/065011.

[42] Gu, Y., Bandopadhyay, S., Chen, C., Ning, C., Guo, Y. (2013). Long-term corrosion inhibition mechanism of microarc oxidation coated AZ31 Mg alloys for biomedical applications, *Mater. Des.*, 46, pp. 66–75, Doi: 10.1016/J.MATDES.2012.09.056.

[43] Wang, C., Shen, J., Zhang, X., Duan, B., Sang, J. (2017). In vitro degradation and cytocompatibility of a silane/Mg(OH)2 composite coating on AZ31 alloy by spin coating, *J. Alloys Compd.*, 714, pp. 186–93, Doi: 10.1016/J.JALLCOM.2017.04.229.

[44] Liu, L., Li, P., Zou, Y., Luo, K., Zhang, F., Zeng, R.-C., Li, S. (2016). In vitro corrosion and antibacterial performance of polysiloxane and poly(acrylic acid)/gentamicin sulfate composite coatings on AZ31 alloy, *Surf. Coatings Technol.*, 291, pp. 7–14, Doi: 10.1016/J.SURFCOAT.2016.02.016.

[45] Kannan, M.B., Liyanaarachchi, S. (2013). Hybrid coating on a magnesium alloy for minimizing the localized degradation for load-bearing biodegradable mini-implant applications, *Mater. Chem. Phys.*, 142(1), pp. 350–4, Doi: 10.1016/J.MATCHEMPHYS.2013.07.028.

[46] Córdoba, L.C., Marques, A., Taryba, M., Coradin, T., Montemor, F. (2018). Hybrid coatings with collagen and chitosan for improved bioactivity of Mg alloys, *Surf. Coatings Technol.*, 341, pp. 103–13, Doi: 10.1016/J.SURFCOAT.2017.08.062.

[47] Yazdani, M., Afshar, A., Mohammadi, N., Paranj, B. (2017). Electrochemical evaluation of AZ 31 magnesium alloy in two simulated biological solutions, *Anti-Corrosion Methods Mater.*, 64(1), pp. 103–8, Doi: 10.1108/ACMM-02-2016-1649.

[48] Alvarez-Lopez, M., Pereda, M.D., del Valle, J.A., Fernandez-Lorenzo, M., Garcia-Alonso, M.C., Ruano, O.A., Escudero, M.L. (2010). Corrosion behaviour of AZ31 magnesium alloy with different grain sizes in simulated biological fluids, *Acta Biomater.*, 6(5), pp. 1763–71, Doi: 10.1016/J.ACTBIO.2009.04.041.

[49] Zomorodian, A., Garcia, M.P., Moura e Silva, T., Fernandes, J.C.S., Fernandes, M.H., Montemor, M.F. (2013). Corrosion resistance of a composite polymeric coating applied on biodegradable AZ31 magnesium alloy, *Acta Biomater.*, 9(10), pp. 8660–70, Doi: 10.1016/j.actbio.2013.02.036.

[50] Song, Y., Shan, D., Chen, R., Zhang, F., Han, E.-H. (2009). Biodegradable behaviors of AZ31 magnesium alloy in simulated body fluid, *Mater. Sci. Eng. C*, 29(3), pp. 1039–45, Doi: 10.1016/J.MSEC.2008.08.026.

[51] Zomorodian, A., Brusciotti, F., Fernandes, A., Carmezim, M.J., Moura e Silva, T., Fernandes, J.C.S., Montemor, M.F. (2012). Anti-corrosion

performance of a new silane coating for corrosion protection of AZ31 magnesium alloy in Hank's solution, *Surf. Coatings Technol.*, 206(21), pp. 4368–75, Doi: 10.1016/J.SURFCOAT.2012.04.061.

[52]   Ye, X., Cai, S., Dou, Y., Xu, G., Huang, K., Ren, M., Wang, X. (2012). Bioactive glass–ceramic coating for enhancing the in vitro corrosion resistance of biodegradable Mg alloy, *Appl. Surf. Sci.*, 259, pp. 799–805, Doi: 10.1016/J. APSUSC.2012.07.127.

[53]   Zhang, Y.F., Hinton, B., Wallace, G., Liu, X., Forsyth, M. (2012). On corrosion behaviour of magnesium alloy AZ31 in simulated body fluids and influence of ionic liquid pretreatments, *Corros. Eng. Sci. Technol.*, 47(5), pp. 374–82, Doi: 10.1179/ 1743278212Y.0000000032.

[54]   Zhang, Y., Liu, X., Jamali, S.S., Hinton, B.R.W., Moulton, S.E., Wallace, G.G., Forsyth, M. (2016). The effect of treatment time on the ionic liquid surface film formation: Promising surface coating for Mg alloy AZ31, *Surf. Coatings Technol.*, 296, pp. 192–202, Doi: 10.1016/J.SURFCOAT.2016.04.038.

Appendix B: *In Vitro* Biocompatibility Assessment

The *in vitro* assessment of a potential biomaterial's biocompatibility is an important initial step for determining the potential safety of a material for implantation *in vivo*. There are two main techniques for assessing the *in vitro* biocompatibility of magnesium, depending on the applications of the device. For cardiovascular applications, such as stents, the compatibility of magnesium is mainly assessed by hemolytic assays (according to ISO 10993-4 [1]) and to a lesser extent by coagulation and platelet aggregation [2]. Hemolysis is the rupturing (lysis) of red blood cells (erythrocytes) and the release of their contents (cytoplasm) into the surrounding fluid. In a hemolysis assay, human red blood cells and test materials are co-incubated in buffers at defined pH values that mimic the desired environment. Following a centrifugation step to pellet intact red blood cells, the amount of hemoglobin released into the medium is spectrophotometrically measured (at 405 nm wavelength for achieving the best dynamic range). The percent of red blood cell disruption is then quantified relative to positive and negative control samples lysed with deionized water and normal saline solution, respectively [3,4]. A hemolytic ratio higher than 5% is considered cytotoxic [4]. For Mg-based materials in orthopedic applications, the *in vitro* analysis of biocompatibility is carried out using cell cultures [5]. Cell culture studies are usually carried out by using one of the two methods adapted from ISO 10993-5 and 10993-12 [6,7]. One method represents an indirect assay in which cultured cells are exposed to an extract of the material of interest. The extracts are prepared by immersing the material in an applicable cell culture media for variable lengths of time. The viability of the cells exposed to the extracts can be measured using colorimetric viability assays to assess the cytotoxic or proliferative effects of the materials. Theoretically, this allows the quantification of the number of living cells if appropriate controls are used. An issue associated with this indirect method is related to the by-products of Mg corrosion that leads to a toxic osmolarity and pH increase that would result in minimal viability. Cell cycle is in fact widely reported to be influenced by the osmolarity; cell proliferation is promoted by osmotic swelling, whereas delayed in hyper-osmotic solutions [8]. Wong et al. [9] reported that a magnesium ion concentration of 50 ppm could stimulate osteogenic differentiation, whereas downregulation of osteogenesis-related genes was observed at a concentration of 200 ppm. This is often avoided by diluting the extraction media before its application to cells, or by increasing the solution volume compared to the size of the sample, with both procedures effectively rendering the same result. The second method adapted from an international standard is a technique involving the direct contact between the material of interest and the cultured cells. Most commonly, this involves the growth of cells directly on the material and is often used to assess cellular viability. It is widely reported that a reduction of cell viability by more than 30% is considered cytotoxic [10]. Further, it is commonly reported that reduced cell viability is due to the change in pH [11]. If the material is being investigated for orthopedic use, osteoblast cell lines or primary bone marrow stromal cells are commonly used [12,13].

In addition, other procedures such as the percentage of survived cells [14,15], antimicrobial activity [16,17], cell density [18–21] and cell proliferation and differentiation [20,22,23] are used to assess the biocompatibility.

REFERENCES

[1] ISO 10993-4:2002 – Biological evaluation of medical devices – Part 4: Selection of tests for interactions with blood.

[2] Liu, X.L., Zhou, W.R., Wu, Y.H., Cheng, Y., Zheng, Y.F. (2013). Effect of sterilization process on surface characteristics and biocompatibility of pure Mg and MgCa alloys, *Mater. Sci. Eng. C*, 33(7), pp. 4144–54, Doi: 10.1016/J.MSEC.2013.06.004.

[3] Evans, B.C., Nelson, C.E., Yu, S.S., Beavers, K.R., Kim, A.J., Li, H., Nelson, H. M., Giorgio, T.D., Duvall, C.L. (2013). Ex vivo red blood cell hemolysis assay for the evaluation of pH-responsive endosomolytic agents for cytosolic delivery of biomacromolecular drugs, *J. Vis. Exp.*, (73), pp. e50166, Doi: 10.3791/50166.

[4] Li, B., Chen, Y., Huang, W., Yang, W., Yin, X., Liu, Y. (2016). *In-vitro* degradation, cytocompatibility and hemolysis tests of CaF2 doped TiO2 -SiO2 composite coating on AZ31 alloy, *Appl. Surf. Sci.*, 382, pp. 268–79, Doi: 10.1016/j.apsusc.2016.04.141.

[5] Lorenz, C., Brunner, J.G., Kollmannsberger, P., Jaafar, L., Fabry, B., Virtanen, S. (2009). Effect of surface pre-treatments on biocompatibility of magnesium, *Acta Biomater.*, 5(7), pp. 2783–9, Doi: 10.1016/J.ACTBIO.2009.04.018.

[6] ISO 10993-5:2009 – Biological evaluation of medical devices – Part 5: Tests for *in-vitro* cytotoxicity.

[7] ISO 10993-12:2012 – Biological evaluation of medical devices – Part 12: Sample preparation and reference materials.

[8] Lang, F., Föller, M., Lang, K., Lang, P., Ritter, M., Vereninov, A., Szabo, I., Huber, S.M., Gulbins, E. (2007). Cell volume regulatory ion channels in cell proliferation and cell death, *Methods Enzymol.*, 428, pp. 209–25, Doi: 10.1016/S0076-6879(07)28011-5.

[9] Wong, H.M., Wu, S., Chu, P.K., Cheng, S.H., Luk, K.D.K., Cheung, K.M. C., Yeung, K.W.K. (2013). Low-modulus Mg/PCL hybrid bone substitute for osteoporotic fracture fixation, *Biomaterials*, 34(29), pp. 7016–32, Doi: 10.1016/j.biomaterials.2013.05.062.

[10] Li, N., Zheng, Y. (2013). Novel magnesium alloys developed for biomedical application: A review, *J. Mater. Sci. Technol.*, 29(6), pp. 489–502, Doi: 10.1016/J.JMST.2013.02.005.

[11] Gu, X., Zhou, W., Zheng, Y., Dong, L., Xi, Y., Chai, D. (2010). Microstructure, mechanical property, bio-corrosion and cytotoxicity evaluations of Mg/HA composites, *Mater. Sci. Eng. C*, 30(6), pp. 827–32, Doi: 10.1016/J.MSEC.2010.03.016.

[12] Walker, J., Shadanbaz, S., Woodfield, T.B.F., Staiger, M.P., Dias, G.J. (2014). Magnesium biomaterials for orthopedic application: A review from a biological perspective, *J. Biomed. Mater. Res. B. Appl. Biomater.*, 102(6), pp. 1316–31, Doi: 10.1002/jbm.b.33113.

[13] Jiang, W., Cipriano, A.F., Tian, Q., Zhang, C., Lopez, M., Sallee, A., Lin, A., Cortez Alcaraz, M.C., Wu, Y., Zheng, Y., Liu, H. (2018). *In-vitro* evaluation of MgSr and MgCaSr alloys via direct culture with bone marrow derived mesenchymal stem cells, *Acta Biomater.*, Doi: 10.1016/J.ACTBIO.2018.03.049.

[14] Pompa, L., Rahman, Z.U., Munoz, E., Haider, W. (2015). Surface characterization and cytotoxicity response of biodegradable magnesium alloys, *Mater. Sci. Eng. C*, 49, pp. 761–68, Doi: 10.1016/J.MSEC.2015.01.017.

[15] Liu, C.-N., Böke, F., Gebhard, M., Devi, A., Fischer, H., Keller, A., Grundmeier, G. (2018). Ultrasound-mediated deposition and cytocompatibility of apatite-like coatings on magnesium alloys, *Surf. Coatings Technol.*, 345, pp. 167–76, Doi: 10.1016/J.SURFCOAT.2018.03.100.

[16] Cheng, M., Qiao, Y., Wang, Q., Qin, H., Zhang, X., Liu, X. (2016). Dual ions implantation of zirconium and nitrogen into magnesium alloys for enhanced corrosion resistance, antimicrobial activity and biocompatibility, *Colloid. Surf. B Biointerfaces*, 148, pp. 200–10, Doi: 10.1016/j.colsurfb.2016.08.056.

[17] Zhao, Y., Shi, L., Ji, X., Li, J., Han, Z., Li, S., Zeng, R., Zhang, F., Wang, Z. (2018). Corrosion resistance and antibacterial properties of polysiloxane modified layer-by-layer assembled self-healing coating on magnesium alloy, *J. Colloid. Interface Sci.*, 526, pp. 43–50, Doi: 10.1016/J.JCIS.2018.04.071.

[18] Hiromoto, S., Yamazaki, T. (2017). Micromorphological effect of calcium phosphate coating on compatibility of magnesium alloy with osteoblast, *Sci. Technol. Adv. Mater.*, 18(1), pp. 96–109, Doi: 10.1080/14686996.2016.1266238.

[19] Amaravathy, P., Rose, C., Sathiyanarayanan, S., Rajendran, N. (n.d.). Evaluation of *in-vitro* bioactivity and MG63 oesteoblast cell response for TiO2 coated magnesium alloys, *J. Sol-Gel Sci. Technol.*, 64(3), Doi: 10.1007/s10971-012-2904-6.

[20] Zhu, D., Su, Y., Young, M.L., Ma, J., Zheng, Y., Tang, L. (2017). Biological responses and mechanisms of human bone marrow mesenchymal stem cells to Zn and Mg biomaterials, *ACS Appl. Mater. Interfaces*, 9(33), pp. 27453–61, Doi: 10.1021/acsami.7b06654.

[21] Huang, W., Xu, B., Yang, W., Zhang, K., Chen, Y., Yin, X., Liu, Y., Ni, Z., Pei, F. (2017). Corrosion behavior and biocompatibility of hydroxyapatite/magnesium phosphate/zinc phosphate composite coating deposited on AZ31 alloy, *Surf. Coatings Technol.*, 326, pp. 270–80, Doi: 10.1016/J.SURFCOAT.2017.07.066.

[22] Ma, J., Zhao, N., Zhu, D. (2015). Sirolimus-eluting dextran and polyglutamic acid hybrid coatings on AZ31 for stent applications, *J. Biomater. Appl.*, 30(5), pp. 579–88, Doi: 10.1177/0885328215596324.

[23] Agarwal, S., Riffault, M., Hoey, D., Duffy, B., Curtin, J., Jaiswal, S. (2017). Biomimetic hyaluronic acid-lysozyme composite coating on AZ31 Mg alloy with combined antibacterial and osteoinductive activities, *ACS Biomater. Sci. Eng.*, 3(12), pp. 3244–53, Doi: 10.1021/acsbiomaterials.7b00527.

Index